W9-ALL-180

INDIANS

of

SOUTHERN NEW JERSEY

INDIANS
of
SOUTHERN NEW JERSEY

by

Frank H. Stewart

KENNIKAT PRESS
Port Washington, N. Y./London

MIDDLE ATLANTIC STATES HISTORICAL PUBLICATIONS SERIES NO. 8

INDIANS OF SOUTHERN NEW JERSEY

First published in 1932
Reissued in 1973 by Kennikat Press
Library of Congress Catalog Card No.: 70-186077
ISBN 0-8046-8608-4

Reproduced from a copy of the original publication made
available by the Special Collections Department of the
Rutgers University Library.

Announcement and Introduction

This is the third publication of the Gloucester County Historical Society, made possible by the income of the following Trust Funds with the Woodbury Trust Company:

Eli and Mary Burnett Steward Fund
Thomas W. Synnott Fund
James C. Griscom Fund
County of Gloucester Fund
Hubert Somers Fund
John W. Sparks Fund

Additional trust funds are solicited in order to increase the number of publications. Memorial, Life, Society and Individual memberships are also desired.

The income of the funds mentioned above has been used to pay for "Foraging for Valley Forge by General Anthony Wayne, in Salem and Gloucester Counties," etc., and the "Organization and Minutes of the Gloucester County Court 1686-7 and the Gloucester County Ear Mark Book 1686-1728."

Numerous other pamphlets have also been published by the Society during the past quarter century of its existence.

All persons are privileged to make use of the contents of this pamphlet provided proper credit is given to the sources of information.

One day in June, 1886, when I was a boy about thirteen years old, while hoe harrowing corn on my father's farm, I noticed what seemed to be an elongated oval-shaped, brown, dry leaf. I stopped the horse, picked up the supposed leaf and was amazed to find it was a stone knife. The discovery was an important one in my life in many respects. During the next two years I found many arrow heads, pieces of pottery and one or two tomahawks, which I saved and took to our former home in Sharptown when we left the farm. On my uncle Nathan's farm, across a run there were three circular depressions at the top of the hillside on Salem Creek which tradition, according to my father, said were Indian tepee sites. On my father's farm, where I was born, near the run that emptied into the creek, there was a multitude of broken stones, chips and rejects that indicated the outdoor workshop of an Indian arrow maker of long ago. Between twenty-five and thirty years ago after I had established a business in Philadelphia, I found that all work and no diversion was wearing me down and I looked around for a hobby. I decided that the collection of Indian relics would be beneficial. The next time I went to Sharptown I asked my mother to get out the relics I had found on the farm and was somewhat dismayed to learn that she had given a Methodist preacher his choice of my small collection, including the first stone knife I had ever seen, mentioned above.

I placed an advertisement in a county newspaper and soon had two or three boys buying stone Indian implements and weapons on a commission basis. It is needless to say that in the course of several months I obtained a well assorted lot of arrowheads, celts, tomahawks, banner stones, mortars and pestles, knives, scrapers, etc., all of which I still have. Among the rare specimens are a crude stone pipe, the only one I have ever seen from Southern New Jersey, a moccasin making outfit of six pieces, a banner stone partly drilled by means of a hollow reed, grit sand and water, two clay pipes, a double groove axe, a duplex mortar and a fine pestle.

I equipped a den in my Camden home and had a four-inch plate rail, on all four sides of the room, with the heavy stone pieces placed on the plate rail rack. The arrowheads were pasted on heavy cardboards and suspended under the tomahawks, which must have numbered between sixty and one hundred specimens. I had a considerable number of books on the subject and finally found myself studying primitive man of the whole world and learned that an arrow head of Egypt, China or Siberia was often duplicated in Southern New Jersey. The bow and arrow, the dugout canoe, the bow drill and fire making equipment existed everywhere with mankind in prehistoric times. The ruins of the temples of Persia, India, Egypt and Yucatan depict the arrow, javelin and swastika cross, the last of which was also a pottery ornament of the American Indians of the Southwest.

From the above my friends may naturally suspect that my trend towards local history had its inception in finding the stone knife on Salem Creek in 1886, which is a fact. For the past twenty years I have paid but little attention to Indians, but still feel that it is a worth while subject, especially for boys, because it may lead to other important hobbies or studies as it did in my case. Millions of dollars worth of merchandise sales will not make as many real congenial friends as a few dozen historical documents, a collection of Indian relics, a hundred fine coins, or a thousand good postage stamps.

The Rosetta stone in the British Museum and the Calender stone in Mexico City compete with the Magna Charta and the Declaration of Independence in historical affairs.

I have climbed a shell mound at Mosquito Inlet, Florida, with the same kind of a thrill as I had in Cheops pyramid on the shore of the Nile. I bought a silver bracelet of a Laguna squaw with the same satisfaction that I purchased an ancient Greek coin in Athens. A stone knife may eventually cause one to wander over the globe.

One time I was on a trout fishing expedition in the mountains of Pennsylvania near Lewistown and there saw a deeply worn trail that had not felt the tread of an Indian for over a century, but I could well imag-

4

ine how Shikellemy and his warriors had followed several generations of their forbears up mountainsides and across valleys for hundreds of years. When I dug up one of the tepee sites on my uncle's farm I found a piece of human skull and a small pottery fragment broken in half. I also collected a number of pieces of pottery and forwarded them to the Smithsonian Institute at Washington which should be visited by every person interested in the North American Indians. The publications of the Bureau of Ethnology should be read.

It seems odd that no present-day land title mentions Indian purchases. Both the Dutch and the Swedes bought land of the Indians long before the English, but because of lack of space this publication will treat of the times after the English took possession of New Jersey.

I have called upon and have been freely given assistance by Mr. Charles A. Philhower, the outstanding authority on New Jersey Indians; Miss Bessie B. Warwick, Chairman of the Indian Relic Committee of our Society; the Historical Society of Pennsylvania, and the New Jersey Historical Society, all of which is highly appreciated.

FRANK H. STEWART.
President Gloucester County Historical Society,
Woodbury, New Jersey

Indians of Southern New Jersey

Indians Doomed by Europeans

The "Noble Red Men" of Southern New Jersey who reckoned time by suns, moons, sleeps, etc., and walked softly in moccasins on wooded, pine-needled paths down to the strands of river, bay and ocean, were doomed the moment the Europeans commenced to trade with them and introduced smallpox, yellow fever, influenza, etc., etc.

In exchange for land, peltries and everything else they had for barter, they received hard liquor, duffels, combs, scissors, pipes, knives, hatchets, shoes, etc., and gewgaws such as looking-glasses, beads, jew's-harps and bells. They also received a few guns, match coats, and a few handfuls of gun powder and some bars of lead.

The trusting, peaceable Indians of Southern New Jersey soon found their hunting fields invaded by the white men who enacted laws that curtailed the former liberties of the carefree Indians.

In the rapid passing of the years some Indians, probably imported, found themselves lawful servants of the white settlers. September 28, 1721, almost as soon as The American Weekly Mercury of Philadelphia started publication, we find Thomas Hill, of Salem, advertising for a runaway Indian marked by smallpox. A few years later Benjamin Acton, also of Salem, likewise advertised a reward for the capture of a runaway Indian servant. In the Inventory of Benjamin Bacon of Cohansey, made May 6, 1714, we find an Indian man valued at 11 L 5 s. Jeremiah Bass, who left money, for preaching two sermons a year, to the Rector of St. Mary's Church in Burlington, had an Indian woman servant or slave valued at 30L in 1725.

The inventory of the estate of William Biddle, of Burlington, shows an Indian woman with two children valued at 100 L in 1711.

Joseph Brown, of Salem County, who died in 1711, had an Indian boy valued at 40 L.

Charles Crosweight, who left this world in 1729/30, had two Indian slaves valued at 50 L.

John Horner, who left money to the Quaker Meeting at Chesterfield in 1715, also had an Indian slave.

Samuel Lippincott, whose estate was inventoried in 1721, had two Indian slaves valued at 75 L.

Daniel Rumsey, of Salem County, also had an Indian man slave when he died in 1718.

John Smith, of Ambleberry, Salem County whose estate was inventoried in 1715/16, possessed an Indian slave.

William Trent, Esq., Chief Justice of New Jersey, who died in 1724/5, owned two Indians valued at 80 L. He also had eight "Indian pictures without frames" listed at 8 shillings.

7

Christopher White, of Alloways Creek, who died in 1693, owned four "Indian notices."

An Act was passed at Burlington in 1713 making imported Negro, Indian and Mulatto slaves dutiable. Whether all of these Indians were imported into the province of West New Jersey or not has not been determined by the writer. It is a well-known fact that some of the early navigators and explorers kidnapped Indians and took them to Europe, and it has been said that Samoset, the Indian who walked down between the huts of the Plymouth Colonists, the next Spring after they were built, and exclaimed, "Welcome Englishmen,!" to their amazement, had been kidnapped, taken to England and returned to the New England coast. The investigations of the writer have led him to believe that it would have been a dangerous practice to enslave local Indians and his opinion, is that Indian slaves of New Jersey were imported from distant places, possibly the West Indies and South America.

Where to Find Indian Relics

The vicinity of nearly every branch and every creek of Southern New Jersey furnishes its quota of arrow and spear heads. The shores of the Delaware have untold quantities of net sinkers of roughly grooved natural stones.

Along the banks of the larger creeks and the river shore are habitation sites of the Indians which are best identified by small fragments of pottery entirely unlike the subsequent pottery of the white people. Indian pottery generally had a quantity of small pieces of shells, or broken quartz, mixed with the clay and showed the outside roughness or impression of the grass or wooden network that supported the clay while being baked. Some of the fragments apparently were rolled with corn cobs and, occasionally, an incised or decorated piece is found. When and where one finds pieces of pottery, arrow heads, etc. may be expected nearby.

The Indian Population

No person knows when the Indians first inhabited Southern New Jersey, neither is there any method of figuring the number of Indians before the arrival of Europeans. Judging from the number of arrow heads, one may think there was a large Indian population, but it is probably safer to guess that they were here a long time before Fort Nassau was built in 1623.

The so-called Indian deeds indicate a small number of Indians to be satisfied. The writer's guess is that two thousand Indian men would exceed the total number here in New Jersey south of Trenton in 1664, and that at the beginning of the Revolution there were not as many as two hundred full-blooded Indian men within the locality mentioned.

Our Indians were not of the ferocious type and nearly all of the isolated cases against the memory of the Southern New Jersey Indians were attributable to "fire water." There never was an organized Indian attack hereabouts, and the murder of the men at Fort Nassau mentioned by several historians is mythical. The only instance worthy of any credence is what happened to a small boat crew from Virginia, mentioned by De Vries, that occurred on Count Ernest River, 1632 (Newton Creek), about half a century before it was settled by the Irish Quakers. Thomas Sharp said the Indians were kind and loving. The Church of England missionary at Salem, who had poor luck in converting them, said they were wild and roving.

One Indian at Salem recorded his ear mark; another one in Gloucester County brought suit in the local court against a white man for an overdue merchandise account.

An Indian was put under bond to keep the peace in Gloucester County in 1699. He was characterized as a "free Indian" which tends to show there were some Indian slaves, as is revealed by the early wills of New Jersey. The Indians, like the Negroes, acquired the family names of the white settlers, and this has helped to obscure them in our old records. Occasionally one finds a baptismal record of an Indian girl.

Indian Treaties

The Colonial Legislature of New Jersey appointed Commissioners to meet the Indians at Crosswicks in the Winter of 1756. The grievances of the Indians were presented to the Legislature by the Commissioners in 1757, when an Act was passed to the effect that no Indian should be imprisoned for debt; that intoxicating liquors should not be sold to Indians, and that no traps larger than three and a half pounds should be set, and that all sales and leases not in accordance with the said Act should be void.

Another Act was passed which gave the Commissioners power to appropriate sixteen hundred pounds for purchasing a general release of all Indian claims to New Jersey, one-half of which was to be used in the purchase of an Indian reservation for Indians of the colony living south of the Raritan River. The other half of the appropriation was to be used to settle claims of the back Indians not resident in the province.

A second conference was held at Crosswicks, N. J., in February, 1758, when the Indians presented a long list of claims for lands, some of them at Great Egg Harbor, also Salem County and Cumberland County. One Indian claimed land at Edge Pillock, on the very tract afterward set apart as an Indian reservation, very likely the first one in the United States.

Abraham Lockquees claimed Stuypson's Island near the Delaware River which was sold by Master Thomas, Indian king, to George Hutchinson, July 10, 1694. This was probably Locquees, the Indian who brought suit in 1733 against Benjamin Worthington in the Gloucester County Court.

Governor Bernard, of New Jersey, made a treaty with the back Indians in June, 1758, by means of messengers sent to Teedyescung,

9

which resulted in a conference held at Burlington, August 7, 1758. A Cayuga Indian, acting as spokesman, said the Munseys were women and could not make treaties for themselves, and said it was not agreeable to have a new council fire kindled or the old one removed to the New Jersey side of the Delaware. He made a pretty speech and delivered a black wampum belt representing the Shawanese, Delawares and Mingoians on the Ohio River. He said the Indian Nations would meet at the Forks of the Delaware the "next full moon after this" and would notify the Governor of the particular day.

The next day, August the 8th, 1758, another treaty was made by exchange of belts, which resulted in an assembly at Easton, Pa., October 8th, when the Six Nations, also The Nanticokes, Conoys, Tutelos, Chognots, Chihohockies, alias Delawares and Unamies, headed by Teedyuscung, with a following of sixty men and women and children, also the Munsies or Minisinks, Mawhickons, Wawpings or Pomptons. The total number of Indians of every nation who attended this conference was five hundred and seven men, women and children. Conrad Weiser, Captain Henry Montour, together with the Indians Stephen Calvin, Isaac Stille and Moses Tatamy, acted as interpreters. Governor Denny, of Pennsylvania, was present with his Council as well as Governor Bernard, of New Jersey, and the Commissioners for Indian affairs. The Conference lasted from October 11th until October 26th and is all fully set forth in Smith's History of New Jersey and the Colonial Records of Pennsylvania Vol. 8. The consideration paid by New Jersey, in full settlement, was one thousand Spanish dollars, for deeds which covered all final claims of the Indians of New Jersey.

The Iroquois, or Five Nations, was originally composed of the Mohawks, Oneidas, Onandagas, Cayugas and Senecas. The Tuscaroras from the South joined with them about 1712 and, after that time, the Confederacy was called the Six Nations. They spoke a different language from the Algonkins. The Indians of New Jersey were subjugated and had no right to make treaties on their own account. This will explain the Conference at Easton which was attended by Indians as far away as the Ohio River.

On March 22, 1762, by virtue of the Act passed in the thirty-second year of the reign of King George the Second, Andrew Johnson, Richard Saltar, Charles Read, William Foster and Jacob Foster, or any three of them, were appointed Commissioners to purchase land for the Indian natives south of the Raritan River to live on.

A tract of land was purchased in Evesham, Burlington County, called Brotherton or Edge-pelick, on August 29, the same year of Benjamin Springer and recorded in Liber C, folio 394, Burlington County deeds.

John Brainard was appointed Superintendent and guardian of the Indians at Brotherton.

Tacktaugh on tana ke matcha; whether goest thou.

Tackomen; whence comest thou.

Undoque; yonder (a little away).

Kecko Larense; what is thy name.

Hickole; yonder (further a great way).

Kecko ke hatta; what hast thou.

Matta ne hatta; I have nothing.

Nee hatta; I have.

Cutte hatta; one Buck.

Nonshuta; a Doe.

Hayes; a skin (not drest).

Hay; a Skin Drest.

Tomoque; a Beaver.

Hunnikick; an otter.

Mwes; an Elk.

Mack; a Bear.

✓ Hoccus; a fox.

Nahaunum; a Raccoon.

Sinqwes; a Wildcat.

Hannick; a Squirrell.

Tumaummas; a Hare.

Miningus; a Minke.

Jwse?; meat, or flesh.

Kush-kush; a hog.

Copy; a horse.

Ninneungus; a mare.

Muse; a cow.

Nonackon; milk.

Makees; a sheep.

Minne; drink or ale.

Pishbee; small beer.

Hosequen; corn.

✓ Pone; bread.

Hoppenas; Turneps.

Seckha; Salt.

Matta; No.

Kee Ningenum une; dost you like.

Me Matta Wingenun; I do not care for it.

Singkoatum; I do not care, I will cast it away.

Mochee; ay or yes.

Me Mauholum; I will buy it.

Ke manniskin une; Wilt thou sell this.

Keeko; what.

Keeko gull une; how many gilders for this.

Kako meele; what wilt thou give for this.

11

Cutte wickan cake; one fathom of Wampum.
Nee Meele; I will give thee.
Cutte steepa; one stiver or farthing.
Cutte gull; one gilder or 6 pence.
Momolicomum; I will lay this in pawn.
Singa ke natunum; when wilt thou fetch it.
Singa ke petta; when wilt thou bring it.
Necka couwin; after 3 sleeps, three daies hence.
Tana ke natunum; when wilt thou fetch it.
Undoque; yonder.
Singa; when.
Jucka, or kisquieka; today, this day, a day.
Hapitcha; by & by.
Aloppau; to morrow.
Tana hatta; whence hadst thou it.
Quash matta diecon; why didst lend it.
Kacko papa; what hast thou brought.
Cuttas quingquing; six ducks.
Olet; it's good.
Matta olet; it's bad.
Matta ruti; it's good for nothing.
Husko seeka; it's very handsome.
Husko Matit; it's very ugly.
Ke runa matauka; thou wilt fight.
Jough matcha; get thee gone.
Undoque matape; sit yonder.
Ne Mathit wingenum; we will be quiet.
Noa; come hither or come back.
Payo; to come.
Match poh; he is come or coming.
Raamunga; within.
Cochmink; without.
Tungshena; open.
Poha; shut
Scunda; the door.
Ke cakeus; thou art drunk.
Opposicon; beyond thyself.
Husko opposicon; much drink.
Mockerick; a great deal.
Tonktis; a little.
Maleema cocko; give me something.
Abij or bee; water.
Minatau; a little cup to drink in.
Mitchen; victuals.
Mets; eat.
Poneto; let it alone.
Husko lallacutta; I am very angry.

Ke Husko nalun; thou are very idle.

Chingo ke matcha; when wilt thou go.

Mesicksoy; make haste.

Shamahala; run.

Husko taquatse; it's very cold.

Ne dogwatcha; I am very cold, I freeze.

Whinna; snow or hail.

Ahelea coon hatta; have abundance of snow, hail, ice.

Take; freez.

Suckolan tisquicka; a rainy day.

Roan; winter.

Sickquim; the spring.

Nippinge; ye summer.

Tacockquo; the fall.

Tano ke wigwham; where is thy house.

Hockung Kethaning; up ye river.

Tana matcha ana; where goes ye path.

Jough undoque; go yonder.

Hitock; a tree.

Hitock nepa; there stands a tree.

Mamanticikan, mama dowickon, mana dickon; a peach or cherry.

Virum; grapes.

Acotetha; an apple.

Hosquen; corn.

Cohockon; a mill.

Rocat; flour or meal.

Keen hammon; grind it.

Nutas; a bag or basket.

Poquehero; it's broke.

Roanouhheen; a north west wind.

Rutte hock; ye ground will burn and be destroyed.

Hockeung; a chamber.

Quequera; where I.

Qulam tanansi; I look for a place to lie down.

Oke cowin; & sleep.

Kee Catungo; for I am sleepy.

Aloppau; to morrow.

Ne nattunum kwissi; I will go hunting.

Takene; in ye woods.

Attoon attonamon; going to look a buck.

Matcha pauluppa shuta; I have catcht a buck.

Accoke; a snake.

Mockerick accoke; a rattle snake.

Husko purso; very sick or near death.

Innamanden; a sore, a burnt, a cut or bruise.

Tospahala; ye small pox.

Nupane; ye ague.

Singuape; hold they tongue.

Singuape hock in hatta; be quiet, the earth has them, they are dead.

Sheek; the grass or any green herb.

Hocking; the ground.

Hockehocken; a plantation.

Nee tutona; my country.

Ouritta; a plain, even, smooth.

Oana; path or highway.

Singa mantauke; when we fight.

Ne holock; do hurt.

Ne rune husev hwissase; we are afraid.

Opche hwissase; always afraid.

Ne olocko teen; we run into holes.

Kocko ke lunse une; what doest thou call this.

Checonck; a looking glass.

Powatahan; a pair of bellows.

Heohoho; a cradle.

Mamolehickon; a book or paper.

Seckock; a table, a chest, a chair.

Sepussing; a creek.

Kitthanning; a river.

Moholo; a canoe.

Rena moholo; a great boat or ship.

Taune ke hatta; where hast thou it.

Ne taulle ke rune; I will tell thee.

Ne maugholame; I bought it.

Ke kamuta; thou hast stole it.

Matta ne hamuta; no I did not steal it.

Taune mangholame; where didst thou buy it.

A B Undoque; yond of A. B.

  B. C. Sicko melee; B. C. will give me so much for it.

Saw we; all.

Hockung Tappin; God (the heavens &c)

Manitto; ye devil (God).

Renus Leno; a man.

Peray; a lad.

Penaltit; a boy.

Issimus; a brother.

Runcassis; a cousin.

Nitthurus; a husband.

Squaw; a wife.

Noeck; a father.

Anna; a mother.

Haxis; an old woman.

Aquittit; a little girl.

Kins Kisite; a maid ripe for marriage

Papouse; a sucking child.

14

Munockon; a woman.
Qualis; a master.
Tollemuse; a servant.
Wheel; a head.
Meelha; the hair.
Skinck; the eyes.
Hickywat; ye nose.
Turr; the mouth.
Wippit; the teeth.
Pentor; the ear or hearing.
Quaquangan; the neck.
Nacking; the hand.
Ponacka; two hands.
Huckun; the back.
Wotigh; the belly.
Hickott; the legs.
Ceet; the feet.
Jucka; a day.
Kishquecon; a week.
Kisho; a month.
Cothtingo; a year.
Passica Catton; a half year.

The Indian Interpreter of 1684

Cutte; one.
Nisha; two.
Necca; three.
Neuwa; four.
Palenah; five.
Cuttas; six.
Neshas; seven.
Haas; eight.
Pescunk; nine.
Tellen; ten.
Tellen oak Cutte; 11.
    "   " Nisha; 12.
    "   " Necca; 13.
    "   " Newwa; 14.
    "   " pallenah; 15.
    "   " cuttas; 16.
    "   " neshas; 17.
    "   " Hass; 18.
    "   " Pescunk; 19.
Nissinach; 20.
Sickenom; a Turkey.
Kahake; a Goose.
Quing-quing; a Duck.

15

Neckaleckas; a hen.
Copohan; a sturgeon.
✓Hamo; a shad.
Huissamech; a catt fish.
Sehacomech; an elle.
Cakickan; a pearch.
✓Lamiss; fish.
Weeko; Tallow or suet.
Pomee; grease or any fat.
Kee-mauholume; wilt thou buy.
Nee mauholum; I will buy.
Kicko-Kee-wingenum; what wilt thou have, or what hast thou a mind
    to have.
Kec-loe Keckoe kee Wingenum; say what thou hast a mind to.
Newing ennun; I have a mind to.
Kake or Sewan; wompum.
Alloquepeper; a hat or cap.
Aquewan; a coat & woolen cloath.
Wepeck a quewan; a white matchcoat.
Limbiss; lynnen cloath or a shirt.
Saccutackan; breeches.
Cockoon; stockings.
Seppock; shoes.
Piakickan; a gun.
Punck; powder.
Alunse; lead.
Assin; stone, iron, Brass &c.
Assinnus; a kettle or pot.
Tomahickan; an ax.
Quippeleno; a hoe.
Pocksucan; a knife.
Tocosheta; a pair of sissers.
Shauta; tobacco.
Hapockon; a pipe.
Brandy wyne; rum &c.
Salem Surveys 2, page 64-68.

John Ladd's Account with Jeremy the Indian

Memo: January ye 9th 1726

Then agreed with Jeremy for halfe a year for ye sum of eight pounds.
Acct of Jeremys lost time since ye 9 of January [11th Month] 1726.

m	th	
11	30	To 1 Day at Whitebrooks drinking
11	31	1 Day at home
12	3	1 Day at home
12	8	1 Day gunning

12	9	1 Day at Tatums drinking
12	10	1 Day hunting deer
		6 Days with ye Indians
		4 Days since
	1727	
2	11	2 Days sick
3	17	7 Days at ye fair
at harvest		2 weeks and 3 days
5	18	8 Days a hunting, Came 26th July
		1 Day sick at ye meadow

June ye 15 1727 Then Jeremy went away to go to harvest work.
August ye 20th Then Jeremy went away

Jeremy the Indian

mo	th	1726			Dr.		
			£		s		d
7	16	To balance due	1	-	1	-	11
		2 lb lead	0	-	1	-	0
		To pipes	0	-	0	-	1
		To 1 pair of garters	0	-	1	-	0
		To 1 lb shot	0	-	0	-	6
		To 1 pair Stockins	0	-	8	-	0
		To 1 pair Do	0	-	4	-	0
		To 1 lb shot	0	-	0	-	7
		To 3 lb Do	0	-	1	-	9
		To cash	0	-	1	-	1
		To 1 shirt	0	-	6	-	2
		To 1 pair buckles	0	-	0	-	10
		To ½ loaf bread	0	-	0	-	3
		To 1 pair garters	0	-	1	-	0
		To 1 lb powder	0	-	3	-	0
		To 1 lb Do	0	-	3	-	0
		To making pair Draws & tape	0	-	1	-	10
		To a shirt	0	-	7	-	6
		To an old shirt & Draws	0	-	3	-	0
		To 1 hat	0	-	4	-	2
		To Cash at Christmas	0	-	1	-	6
		To 1 hat	0	-	6	-	4
mo		To an old Jacket	0	-	14	-	0
11	19	To cash pd to Michl Lican	0	-	6	-	0
11	19	To cash paid I. Butterworth	0	-	1	-	6
11	28	To cash to go to Gloster with	0	-	1	-	0
12	8	To ½ lb powder	0	-	1	-	6
		To 1 lb shot	0	-	0	-	7
		To 1 lb powder	0	-	3	-	0
1	3	To ½ lb powder	0	-	1	-	6
		To 1 lb shot	0	-	0	-	7

		To 1 lb shot for ye Indian boy	0 - 0 - 7
2	20	To 1 felt hat of James New	0 - 6 - 4
		To 8½ yards of checkt linnen	0 - 17 - 0
		To 3 yards of lining @ 2/4	0 - 7 - 0
		to Indian Woman	
		To 4 yards & Qur of ditto	0 - 10 - 7
		To making a shirt & thread	0 - 2 - 6
		To 3 yards, ozinbrigs	0 - 4 - 6
		To making 1 pair Trowsers	0 - 1 - 6
		To making 2 speckled shirts & thread	0 - 3 - 0
		To a pair of shoes of S. Warner	0 - 7 - 6
		To a pair of stocking of Trapnell	0 - 6 - 6
		To 2 lb tobacco	0 - 0 - 8
3	18	To cash to go to ye fair	0 - 2 - 6
5	24	To Qur of powder	0 - 0 - 9

1726		Contra	Cr
			£
7 mo	17	by a turkie	0 - 0 - 6
7 "	20	by 2 Qurs venison	0 - 1 - 6
		by 2 deer	0 - 7 - 0
		by 1 fawn	0 - 2 - 0
		by 1 Do	0 - 2 - 0
		by 1 goose	0 - 0 - 6
		by 24 ducks	0 - 4 - 0
		by 2 Qurs of a fawn	0 - 1 - 0
		by 18 days work for John Trapnell	1 - 7 - 0
		by 3 days work	0 - 4 - 6
		by 5 ducks	0 - 1 - 3

2 - 11 - 3

At this time March was the first month and February the twelfth month. The year began on March 25th or New Year's Day.

The above account is taken from a book kept by John Ladd, Jr., now in the possession of the compiler. While Jeremy may have been a trifling workman he was unquestionably an experienced hunter and good marksman. The taverns, the fair, the Indian woman, the wild turkey, etc., make one think he would like to see Jeremy the Indian of Old Gloucester County where John Ladd, Jr., gathered most of the political plums for half a century.

Indian Trails

Mrs. Blackman wrote that one Indian trail started at Somers Point and extended along the east side of Great Egg Harbor River so as to go to the north of the heads of the several branches of Babcock's Creek, and over the lowlands made by the near approach of that branch to some of the tributaries of Little Egg Harbor River called "The Locks" by Blue Anchor and crossing the head of Great Egg Harbor River at Long-a-coming (Berlin) passing a short distance south of Haddonfield and striking the Delaware at Cooper's Point. Another trail started at the mouth of the Little Egg Harbor River in a westerly direction and joined with the other trail near the head of Landing Creek, one of the branches of the said river. The third trail began near Mullica's plantation a short distance from Batsto, went in a westerly direction between the streams of the first mentioned trail at the old Beebe place about one mile south of Winslow. The fourth trail she mentioned as the Old Cape Road, starting in Cape May County. It crossed the head of Tuckahoe River in a northerly direction, and to the west of the branches of Great Egg Harbor River to the upper waters of Hospitality Stream at Coles Mill, thence to Inskeep's ford and joined the first named road at Blue Anchor.

"A Going - Over Place"

In a deed dated Elsinboro, Salem County, July 14, 1794, signed by Clement and Hannah Acton, his wife, to Clement Hall, mention is made of the "going-over place" a little below the head of "Mannamuskee" creek. The deed is witnessed by Edward Hall and Joseph Sloan and appears to be unrecorded. The writer has always felt that the "going-over" places went back to our aboriginal people with moccasined feet.

Indian Path Glimpsed

In a deed (unrecorded), dated April 30, 1759, of Sam'l Harrison and Abigail, his wife, to Joseph Ellis for one lott of Cedar Swamp situated on the easterly side of Great Egg Harbor River in the County of Gloucester at the mouth of a small branch next below the placed called the Indian path and is bounded as followeth: Said ten acres of cedar swamp is part and parcel of a survey made by Samuel Clement Jr. Sept 14 1758 entered on record in the Surveyor Generals Office in Burlington in Liber H Folio 166.

Leah Blackman's History

Leah Blackman, in her history of Little Egg Harbor Township, devotes several pages to the Indians of that township, among them "Queen" Bathsheba Moolis. Mrs. Blackman stated that when the Edgepelick Indians removed to Lake Oneida in New York State in 1802 the most of the small remnant of the Little Egg Harbor tribe went with them where they remained until 1824 when they removed to the Indian purchase on Green Bay. She gives the Indian numbers up to twenty, evidently obtained locally at or near Tuckerton.

During the Centennial year two or more of the Western Indians visited Indian Mills, the ancestral home, according to a man living there.

Indians Killed Settlers' Hogs with Arrows

From a mildewed piece of paper, now ruined, the writer obtained the following a few years ago:

On the 1st of 4th month 1689 an unknown settler petitioned the Justices, sitting in Gloucester County, to the effect that he had sustained and suffered great damage, by the Indians, who had destroyed several of his hogs, and that one Indian had confessed killing several of them. He had obtained no satisfaction from the grand jury to whom he had evidently appealed. "Five great brooding sows with several hoggs more were wanting" and the unknown petitioner had reason to believe the Indians had killed them for they had often "come home bloody being shot with arrowes." The name of Tammehauck appeared and he possibly was the Indian who had admitted the slaughter. The petitioner claimed he would be ruined if not sustained in his claim. On the other side of the powdered piece of paper this appeared: "Gloucester ye 5th of 7th month 1689. The Bench awards that the goods, formerly delivered by ye Indians, for satisfaction of Hoggs killed by them of . . . be brought to one place viz., [John] Reading and their appraised by ye . . . impartial judgments, of Thomas Thackara and William Alberson, County . . . who shall make report of their proceedings for ye next court, to be held at Gloucester; the appraisers to meet upon ye 7th of yt Instant at ye place abovesaid. The division to be made in manner following; first thirty-seven shillings to be paid for ye Hoggs then disburst . . . and fees and . . . allowance . . . of tyme."

The county appraisers seemingly divided and disbursed the goods obtained from the Indians who killed the hoggs. The writer was unable to find anything in the court records pertaining to the case.

Gloucester County Court Records

"This to certifie the Court at Gloucester that Callouque-hickon an Indian brought to my house a painter [panther] head the 15th of the 2 month [April] 1692-3 and was by me ordered to be or send to the next Court to require pay twelve guilders for killing the same.

John Wood"

5th 7 mo [Sept.] 1689 The Grand Jury with ye concurrence of ye Bench agree and order that whoever shall pay for any wolfs head to any Indian or other person shall be repaid by ye Treasurer of ye County with interest for his money so disbursed.

On September 1, 1690 Peter Matson a witness in a hog-killing case testified he was up Great Mantua Creek about the first of April on account of burying an Indian child.

Unwanted Indians

An unsigned copy of a petition to William Franklin, Esq., Captain-General, etc., shows that certain Indians, to the number of twenty-one belonging to some tribe in the Colony of Virginia or Maryland, had been lately taken under the protection of the Government of Pennsylvania and had for some secret reason been sent over to Gloucester County, under the pretense of having them again removed within a few days.

This not being done as agreed the petitioners requested that the Governor cause it to be accomplished. The writer thinks the printed archives of Pennsylvania and New Jersey will shed some light on the subject.

A Law of the Indians

Governor Franklin was well aware of the unwritten law of the Indians. If a white man killed an Indian and no financial settlement were made and the culprit escaped unpunished, a white person's scalp was likely to be taken in retribution. The backland Indians have been known to travel several hundred miles to seek revenge for a murder, of one of their nation, by a white man. The Indians were prone to kill innocent settlers in the application of their laws. This feature had a great deal to do with the warfare and massacres between the Indians and whites on the frontiers. After Seventeen hundred and fifty the Indians were widely scattered but they always retained an interest, if not love, for their homelands and their relatives who remained thereabouts.

Indian Names

The writer is indebted to the New Jersey Historical Society for a list of Indian names of places and streams in the lower counties of New Jersey. In Atlantic County are Nacote Creek, (Niskeute) nasty, dirty; Neshochcaque, (Niskewack) a muddy place. Other Indian names in the list of Burlington, Camden, Cape May, Cumberland, Gloucester, Mercer, Ocean and Salem Counties: Pensaukin, Rancocus, Assiscunk, Papoose Creek, Tulepehackin, Meekendam, Tuckahoe, Oronickon, Nantuxet, Muskee, Manamuskiss, Menantico, Manaway, Raccoon, Mantua, Nahoney, Assinpink, Shabakunk, Waycake, Manasquan, Metecunk, Matawan, Conaskunk, Westecunk, Manahocking, Congasa, and others, principally names of streams.

To the above may be added Almonéssen, Alloways, Nicomus, Cohawkin, Patcong, Cohansey, Mannington and many others.

Volume twenty-one, Calendar Records, of the New Jersey Archives contains considerable information about Indian deeds for land and the names they applied to their towns and the water courses. Arwamos was the Indian name for the site of Gloucester City. Oneanickon or Hony-Honickon was an Indian town in Burlington County, as was Itchalamensey, and Alumhatta on Rancocus Creek. There was an Indian town on the west side of the main branch of Alloways Creek in Salem County.

James Cooper Griscom, of Woodbury, has an old deed for land at the head of Hesters branch, dated July 2, 1713, from Richard Bull to John Cooper, in which Succotanny, an Indian town, is mentioned. A large spring of water is still known by that name. Hester's branch is a tributary of Woodbury Creek, named for Thomas Hester, and flows across Cooper Street, east of the Railroad station at Woodbury.

King Nummy Found a Whale and Sold It

Burlington Court 4-7 mo, 1685, records of a law suit show that an "Indian called Nummy" sold a whale he had found with a harpoon in it.

21

John Peck, 12-16-3rd mo, 1688, mentioned in a law suit, Burlington County Court. He was a whaler accused of disposing of Dubartis whales on shore, contrary to law.

This man was the one who gave his name to Pecks Beach, now Ocean City. He did not own it. (The writer hears the roar of its surf while writing this.)

An Indian Canoe

In the Spring of 1929 five boys, while prowling around a ditch back of the Methodist Church at Hurffville, found the remains of about five

feet of an Indian dugout canoe. Not realizing its importance they used force in prying it out of the mud and broke off the remaining parts of its sides. The writer was informed that the ditch was a tributary of Mantua Creek but in the course of time had partly filled with mud. The relic was obtained for the Gloucester County Historical Society and was placed in a glass showcase in its museum. The half-tone illustration shows what was left of the canoe at the time it was found. A school girl of Woodbury very kindly held two broken pieces in position while the snapshot was made, on the lawn in front of the Historical Society. At the time of its discovery it was water-soaked and heavy, but when it was thoroughly dried it became light and punky. The bow of the canoe is wedge-shaped and this possibly indicates that it was made so, by use of an iron implement, and shortly after the advent of Europeans, but before sawmills, were on many of our streams, probably before the beginning of the Eighteenth Century.

Shortly after the Mantua Creek canoe was found another very large cedar dugout canoe was located at Dennis Creek and taken to the New Jersey State Museum at Trenton. It had been through a fire but what was left of it, ten or twelve feet long was in almost perfect preservation. This canoe was probably not as old as the smaller one and was undoubtedly made by white men. The crudeness of one and the perfection of the other, as well as the great difference in size, tends to make a person think one was made by Indians and the other by our early settlers. They are both interesting specimens of primitive times in Southern New Jersey.

Odds and Ends

The diaries and account books of Samuel Mickle, of Woodbury, show that he bought baskets of the Indians of Brotherton or Edge-pelick.

John Hunt, in January, 1798, met an Indian chief in Philadelphia who lived beyond Detroit. His interpreter was one Isaac Zane, who had been captured by the Indians when he was nine years old and had lived with them about forty-three years.

One P. Schagen, a Dutchman, in a letter dated November 5, 1626, said that the Indians had been paid sixty guilders for Manhattan Island. He mentioned a ship that sailed with 7246 beaver skins, 675 otter skins, 48 mink, 36 wild cat, 34 rat and other skins of limited number.

Thomas Sharp, a first settler on Newton Creek, recorded on page 98 of deed book A of Gloucester County that the first Irish Quakers were somewhat doubtful of the Indians in 1683.

Missionary John Holbrook, in a letter of 1725 to the Society for the Propagation of the Gospel, wrote "As for the Indians there are very few seen here. They are a wild roving people and never remain long in one place." According to his letter, Salem then had thirty houses, Greenwich about twenty and Fairfield a like number.

Volume One of the New York Colonial Documents, page 281, in a description of wampum says that the hand or fathom was the measurement and that some of the Indian houses had portraits and pictures somewhat crudely carved. This book should be consulted by students. The Freeholders' records of Gloucester County show that an old, destitute, sick Indian had been abandoned by his people about two centuries ago.

Indians were sometimes admitted to the poor houses but the most of them preferred to live in their own shacks. Traditions still exist of them at Sharptown, Cross Keys, Mickleton and Medford.

John Ireson and Richard Whitaker were indicted, in Old Gloucester County, October 1, 1693 for selling rum to the Indians.

The Brotherton Indian reservation contained slightly over 3000 acres.

South Jersey Indians on the Bay, the Cape and Coast

Proceedings New Jersey Historical Society January, 1931

By Charles A. Philhower

Investigation is more and more leading to the conclusion that the American Indian as the earliest explorers found him was not the first inhabitant of the state now known as New Jersey. Certain stone implements found by C. C. Abbott embedded in the Trenton gravels suggested that man in America was much older than history had accredited age to him. The crudity of form of the objects found, the age of patina, depth of position in geological formations, relationship to animal remains, Indian traditions and etching of an ancient character on rocks and stones have suggested that man in New Jersey may date back to a period coexistent with the age of mastodons, mammoths and dinosaurs. This would mean that there were human beings in New Jersey ten to fifteen thousand years ago.

Troglodyte beings may have inhabited the rock shelters of Sussex, Warren and Hunterdon. Paleolithic man may have fought off the dinosaurs from his preserves, and neolithic man might have been responsible for the extinction of the mastodon whose remains are found in every quarter of the State.

There seems to be every reason to believe that man in America is as old as primitive man in Europe. It rests with the scientist of tomorrow to prove this hypothesis set up by Dr. Abbott some 50 years ago.

Described by Early Explorers

The history of the Indians of New Jersey begins with the account given by Verrazzano who touched our shore at New York Harbor early in the Spring of 1524. He states, "We found the country on its banks well peopled, the inhabitants not differing much from the others to the South, being dressed out with the feathers of birds of various colors. They came towards us with evident delight raising shouts of admiration."

On his third voyage Hudson entered Delaware Bay, August 28, 1609, September 4th he discovered the Raritan River and Robert Juet, in the log of the Half Moon, gives an account of the natives found on the shore of the river. An excerpt from his entries reads as follows: "This day the people of the Country came aboord of us, seeming very glad of our coming and brought greene Tobacco, and gave us of it for Knives and Beads. They go in Deere skins loose, well dressed. They have yellow Copper. They desire Cloathes, and are very civill. They have great store of Maiz or Indian Wheat, whereof they make good Bread."

On the fifth he writes: "This day many people came aboord, some with Mantles of Feathers, and some in skinnes of divers sorts of good Furrs. Some women also came to us with Hempe. They had red Copper Tobacco pipes, and other things of copper they did weare about their neckes. Soon they went on Land againe, so wee rode very quiet, but durst not trust them." On September 6, Hudson entered New York Harbor.

Captain Samuell Argall on August 27, 1610 sailed up the Delaware Bay in the vessel Discovery and "found great store of people which were very kind, and promised that the next day in the morning they would bring him great store of Corn." However, the wind blew him out of the river and he saw no more of the Indians. This was the first vessel to anchor in the River. It was Argall's pleasure to name the river for his friend, Sir Thomas West (Lord de la War) who had come to the Virginia Colony with him but stayed at Jamestown while the expedition went out from that point.

Captain Cornelis Jacobsen May in the ship Fortune entered the river on October 11, 1614.

On December 3, 1632, David Petersz De Vries sailed up the river in his yacht The Squirrel. He gives a detailed account of the Indians about Manhattan but makes no mention of the natives along the Delaware.

It was left to Peter Martensson Lindestrom, a student of the University of Upsala, at Stockholm, to give the world the earliest description of the natives living in the Delaware Valley. On the Swedish Government's tenth expedition to New Sweden, Lindestrom was granted passage with the understanding that he would aid in fortifying the Delaware and in mapping the river. He arrived in the Fall of 1653. It was in his manuscript entitled, "Geographia Americae," that he described the natives. This great source book was translated into English in 1925 by Amandus Johnson and is an invaluable piece of work to students of the Lenape Indians commonly called Delawares.

The Lenape Nation: Its Sub-Tribes and Totems

The Indians that occupied New Jersey during and immediately preceding historic times were known to themselves and by their contemporaries as the Lenape which means "Indian man." On occasion when addressed in an emphatic mood they were hailed as the Lenni Lenape, or "men of men." They lived mainly in and about the Delaware River Valley occupying the country westward to the divide separating the Delaware and Susquehanna watersheds. The Chesapeake peninsula lay within their domain and the Hudson Valley South of the Catskills was Lenape territory. Long Island and Staten Island were occupied by the Delawares and to a degree the eastern shore of the Hudson north to the region of Dutchess County was claimed as their hunting ground.

The Delaware River was known to them as the Lenape-wihituk, the domain of the Lenape. It is said that their ancient council-fire was at Shakamaxon, meaning the place of eels, located near Kensington, Phila-

25

delphia. The Hudson was called the Moheganwihituk, the territory of the Mohegans. The Lenape name for the Atlantic Coast, Zeewanhacky, shows decided Dutch influence. They took the Dutch words, zee, meaning sea, and wan, meaning fan and combined them with their locative ending, hacky, making the word Zeewanhacky, signifying the place of sea fans or shells. New Jersey was called Schejachbi, which is derived from scheyek, meaning a string of wampum, and bi meaning water and is interpreted, the land of shell wampum.

The Lenape were the dominant group of a triple alliance including the Mohegans of the Southern Hudson and Long Island and the Nanticokes of the State of Delaware and the Chesapeake Peninsula. They were comprised geographically of three sub-tribes. The Minsi occupied the region extending from the Catskills in New York to the Musconetcong and Watchung Mountains in New Jersey. Their totem was the wolf, tookseat in the Minsi dialect, meaning round foot. Their council fire was situated at the Great Minisink Island in the Delaware River, twelve miles south of Port Jervis, New York. This sub-tribe served as a buffer state between the Lenape nation and the Iroquois.

The Unami occupied the Delaware Valley south of the Minsi. Their totem was the turtle, pako an go, the crawler. Their council fire was at Trenton, chickihoke, meaning, at the tumbling bank. It is said that they always took the lead in civil affairs.

The third sub-tribe, in whom we are particularly interested, were the Unalachtigo. They were closely related to the Nanticokes. In fact, lachtigo and nantico are cognate words. The cultures in South Jersey and the Chesapeake Peninsula have many common characteristics. The crescent ceremonial stone, the small polished celt, the large wooden mortor, the eel fike, the spike-tongued wooden net needle, and the round-based arrowhead are implements frequently met with in both areas. The traditions of remnants of these tribes living in South Jersey and the Peninsula point to a common ancestry. Furthermore, their common enemy was the Susquehannocks with whom they exercised a deadly feud.

Relation to Susquehannocks

The Susquehannocks, who were doubtlessly the giant-like race which the Lenape encountered east of the Mississippi, continued up until their extermination to be the enemies of the Lenape. Since in the earliest time the Unalachtigo were apparently the ruling sub-tribe of the nation, their enmity against the Unalachtigo and the Nanticokes seemed to be most bitter and continued until the last. Whether this contention had a bearing on the loss of position of the Unalachtigos raises question. Frequent references to the Unalachtigos being at war with the Susquehannocks are noted. The conflict of 1623-1634 drove the Unalachtigos, who lived on the west shore of the Delaware, down into the Chesapeake Peninsula, and those on the eastern shore to the Cape and the Coast. Burials unearthed on the Gilson Farm at Tuckerton suggest the capture and execution of some of these giants.

It is said that the Iroquois were never directly engaged in warfare with the Lenape but that their attitude of collecting tribute from them came about when the Iroquois conquered the Susquehannock who had previously subdued the dominant sub-tribe, the Unalachtigo Lenapes. After this it appears that the Unamis became the governing tribe of the Lenape.

Physical Characteristics of Unalachtigos

The Unalachtigos were brownish in color, varying in height, some tall, some medium and some short. They were invariably well proportioned, slender and straight as a candle. In feats of physical strength it was observed that they were weak across the back but strong and agile in their limbs. Their training inured them to hardship, enabling them to withstand drastic exposure and to run with as great endurance as a dog in the chase.

The Dress and Attire of the Unalachtigos

In the early accounts of the Indians in South Jersey no mention of baldheadedness is made. They were found having by nature great crops of long, coarse black hair. It was the custom of the braves to cut the hair with sharp flints. Here and there shocks of hair were left so that there were some half-dozen tufts on the head. Each tuft was tied about or braided, and was decorated by threading shell wampum at the end of the lock. Long feathers from the eagle, buzzard and raven, gayly colored, were set in these clumps of hair. The feather head-dress of the Sioux was never found among them. Sometimes shocks of hair about the ears hung down to the knees. Those parts of the scalp devoid of hair were painted red, giving the individual a horribly gruesome appearance. About the forehead both men and women would wear bands of wampum with figures of animals and birds worked in them. Squaws wore their hair long. Sometimes it was very much disheveled. Generally it was braided in four locks and treated with bear's fat. On some occasions the hair hung down the back and on others it was tied up into a pouch.

It was the practice to pull out the hair on the face so that the chin of a man was often as smooth as that of a woman. The men painted their faces black, red and green with lines, circles and figures that gave them a frightful look. The ears were invariably decorated with wampum or copper rings. The wealth of an Indian was ever apparent in ceremony or the dance by the number of strings of wampum that hung from his neck. In among this wampum the totem or paahra of the brave was to be seen. This may be a polished piece of shale or soapstone representing a fish, bird, animal or insect. Frequently this paahra was in the form of a pendant, gorget or ceremonial stone usually plain but often with drawings faintly etched on the surface.

On their backs they carried a bag called a natasson as a receptacle for food, pipe, tobacco, wampum and trinkets. This natasson was often large enough to hold bows and arrows for the chase.

A cloak-like garment of beaver, deer, bear or raccoon skin was worn with the skin inside in warm weather. This dress was reversed in winter. With the men there was a wide belt of wampum about the waist ornamented with figures which served as a girdle.

Dangling from their cloak were many narrow strips of leather strung with beads and wampum. The trunk and limbs were bare in summer excepting in a few cases where bands of copper were worn about the wrists and ankles. A kind of shoe called a sippack covered the feet. It was commonly made of deer skin and was laced up with a thong. Sometimes they were made of corn husks. These were also much decorated with shell money.

Skill and Adaptability

They exercised much skill and adaptability in working with their hands, deftly shaping their stone arrows with naught but a stone maul and a piece of antler. Their resourcefulness with the crude tools at their command astounded the early explorers. It is said with nothing but fire and a stone chisel an Indian would go into the forest and in three to five days produce a complete dug-out canoe made from the trunk of a live white oak, chestnut, poplar or sweet gum. Baskets and mats of white oak, poplar and water ash were much in demand by the settlers. Wampum called gock was expertly wrought from the shells of the quahog and polished most attractively.

Quick to Adopt European Tools, Implements and Cloth

They were quick to imitate the practices of the white man. In shooting with guns and rifles they soon exercised a high degree of skill. Axes, needles, awls, hoes and knives of iron were adopted immediately. The use of scissors, tobacco, bones and tongs seemed to present no difficulties. They quickly grasped the use of combs and sought in trade rings for their fingers and bracelets for wrists, arms and ankles. The beaver skin and elk skins used for centuries as clothing were summarily exchanged for the gay colored Dutch duffel cloth of blue, orange and red. Stockings of woolen supplanted leggings made of skins and match coats and epaulets characterized the sachem on parade.

Fond of Gewgaws

Tinshaw looking glasses appealed to their vanity and the Jew's harp readily found a place in their daily amusement. Bells for the dance and jingling sheets of metal that dangled from their skirts were much sought for articles usually considered in barter and exchange. Copper kettles and pewter spoons were adopted as part of their culinary equipment; but the fact that fingers were made before forks was always evidenced by their practice. One object that often took precedence even over the gay colored beads of glass and highly prized shell wampum was the fish hook. It was an article both of pleas-

28

are and necessity. Clay pipes introduced by the Dutch were commonly used in pastime smoking but the ornamented calumet retained its place on all occasions of ceremony. Gaudy pottery of Dutch manufacture was much admired and though it was soon demolished by use it never failed to attract their fancy. Even a handful of bits of colored glass would buy a beaver skin. The Unalachtigo, though conservative in many of his practices, never failed to respond to the appeal of color, novelty and obvious utility.

Skillful as Trappers and Marksmen

In the field he was uncanny in his attainments. He had learned to outwit the most alert animal of any species whatsoever. The bear was followed to his lair with ease and certainty. The fox was trapped unawares. Wind and weather were taken into account and the lure of bait and animal calls were ever at his command. The trap known as the den type dead fall was his invention and the spring stick snare, and the bear pit always served him to great advantage in filling his larder. He could give the call of the wild turkey, quack with the perfection of a duck and honck as adept as a goose. He could growl like a bear, bark like a fox and squeal like a rabbit in distress; and when within his reach he took the game with bow and arrow, spear, javelin or dagger.

Sports, Training and Development of Boys and Girls

Children were left much to themselves. They would chase each other, run races, throw stones, climb trees, play in the water and fight and tussle about. They were taught early to use the bow and arrow, the spear and javelin. Much time was spent in fishing. They would feel under stones and catch fish with their hands. It was called "fingering fish." Other amusements were hunting bird nests, gathering shells, pebbles and marking stone. They would wander in the woods and locate the homes of wild animals. It was most important that every boy should learn how to make fire with flint stones and cedar bark. Then, too, how to preserve fire and transport it by means of hickory punk was a most important part of his schooling.

When a boy approached manhood there was the longed-for ceremony that made of him a warrior. This was called huskanawing. The youth was taken into the forest by a brave, shut within a latticed enclosure and left without food or drink. His courage was tested by the animals that would lurk about barking, snarling and growling at him. Finally after exhaustion through fear, hunger and thirst his attendant would give him a concoction of jimsonweed which would throw him into a deep sleep. On arousing from his lethargy he would be taken back to camp and the animal most prominent in the story of his dream became his totem to which he looked for success in war, famine and the chase. After appropriate ceremony, at which time a feather was put in his scalp lock he, who had been huskanawed and had withstood this trying ordeal, was at liberty to use the Delaware war cry "Husca n'lenape win"—meaning,

29

truly I am a Lenape. An effigy of the animal was worked out of stone and worn as his paahra or sacred charm.

When a girl came to marriageable age she wrought for herself a beautiful dress of turkey feathers. The appearance of such a dress must be left to the imagination for neither a drawing nor a description is to be found in any of the old accounts. Girls wore an apron from their earliest childhood but the boys ran about naked. Infants were bound to a board for a period of about a year. An apron of beaver skin was worn about the middle with the fur on the inside in winter and with the skin on the inside in summer. Both men and women in cold weather would wrap themselves up in large skins of elk, deer or bear. It will be concluded that the Indian in his dress was not devoid of adornment, protection and comfort.

Language

The language of the Unalachtigo had a comparatively small vocabulary of about 5000 words. One word may have various meanings, for example:

och que'chum—a female animal
och que hel'lan—a female bird
och que'tit—a little girl
och que'u—a woman

In the Minsi and Unami dialects of the Lenape there was no letter (r), but the Unalachtigos were accustomed oftentimes to use an (r) instead of an (l). They said Renape instead of Lenape. A brief conversation will give you the sound of their language.

Kee squa og enychan hatah? Hast thou a wife and children?
Mogy. Yes.
Katcha hatah? How many?
Neo. Four.
Nisha benointid og nisha squatid. Two boys and two girls.

Considered Cunning and Revengeful in Warfare

It is said that when a war arose with another people a Unalachtigo chief would call his people together and change the names of things so that they may not be understood by the enemy. It would seem by this practice that they were very versatile in their speech.

Their ruler was called a sachem or sakima. He lived on a level with his fellows in time of peace; but when the tribe was at war he was looked up to as a leader and his word was law. They practiced the principle of like for like. If the enemy killed one of the tribe, peace could not obtain until another was killed in return. They fought with spears, daggers, and bows and arrows. The war arrowhead was a small triangular flint, the barbs of which would catch when shot into the flesh. One pulling on the shaft would loosen it from the head and the point, poisoned with the venom of the rattlesnake, would be left to do its deadly work. It

30

was for this reason that Cooper, in "The Last of the Mohegans," called the triangular arrowhead the "loose arrowhead."

Indians Poor and Eager to Trade

The Indians in the southern part of the State were found by the Swedes to be very poor. While they never were in need of food because of their proximity to the river and the shore, and even though beaver was plentiful in the swamps and shells for wampum littered the coast, yet they possessed little, if any, wealth. This is explained by the fact that they sought for nothing beyond the necessities of life. Sufficient unto the day set the limit of their needs. The Swedes and Dutch were ambitious to develop trade in beaver peltry. This was brought about in a limited way for the zeewan (the Dutch name of wampum or gock, mean-zee, meaning sea; and wan, fan, sea fan or shell) was much better than that of Indian manufacturer and was ardently sought for as adornment of dress.

The colonists were always in need of corn, fish and venison, and the natives were looked to for furnishing them. To these they added deer meat, wild turkey, the heath hen, swan, geese and ducks. For such commodities they received frieze, duffel cloth, kettles, axes, hoes, knives, fish hooks, mirrors and the like. More desirable commodities such as guns, powder, lead and rum were withheld for the purchase of land.

Foods Plentiful

Suppan or corn meal was the staple of the native. Marrow fat was used as butter. Besides the meat of wild animals, fish and fowl, there was much to be had from the soil. The subterranean fungus called tuckahoe was common in the southern part of the State. The wild cactus or Indian fig is still the index of an old village site. The orange-colored squash, grown throughout the lower reaches of the State, is of Indian origin. Beans and pumpkins were dried for winter use. The beach plum and persimmon were gathered in season and chestnuts, hickory nuts, walnuts, butter nuts, hazel nuts and acorns were stored away to avoid famine. Fruits such as huckleberries, cranberries, blackberries, raspberries and strawberries gave variety to their fare.

Most everywhere on the coast, and on water courses, within tide water and on the bay, the soil is littered with shells of the oyster, clam, scallop, snail and welk which are telltale evidence of native feasts. Mixed with these are shells of sea turtles and the bones of deer and bear broken for the marrow fat.

Preparation of Foods

While the Indians ate meat and fruits raw, yet there was a general practice of cooking their meals. This was done over the open fire. Portions of meat were fastened on the end of a spit and broiled in the flames. Pots were made of clay for cooking meat and vegetables. These ves-

31

sels had egg-shaped bases, were set in the sand and a fire built about them. The pot with a constricted neck for suspension appears never to have been used in the southern part of the State. Cakes were made of cornmeal ground by means of the stone pestle and mortar. A fire was built on a bed of stones and, after these were made hot, the embers were swept away with a split broom and small loaves wrapped in the green leaves of the sassafras, tulip poplar, oak or chestnut often pinned together with thorns, were placed on the hearth to cook. Mixed with these cakes were dried berries, plums or chestnut meal and when well cooked they were very palatable. Probably the most common dish was the bowl of suppan or mush. This was cornmeal boiled in water and served in wooden bowls. Oftentimes each member of the family with a shell would dip the porridge from the pot alternately while it was simmering over the fire. Fish, meat and shell fish were dried and smoked for winter use. Squash was cut in strips and hung up to dry. Corn and nuts were charred and stored in the ground in large clay pots. Roots and tubers such as the ground nut, artichoke, Indian turnip (the bulb of Jack-in-the-pulpit) Solomon's seal, calamus and the like were gathered in the fall and buried. In certain parts of this southern area garlic was developed to a considerable size and used extensively as a food. It will be noted that through frying, boiling, smoking, charring and baking the Indian squaw prepared her meals in a rather acceptable way and that they did not fare so badly for food after all.

Sickness, Medicine, Death and Burial

In the field of medicines and nostrums their practices were not without results. The medicine man or pow-wow doctor administered his prescription when resorted to. However, the squaws knew much in the use of common remedies. Teas such as bonset, penny royal, and catnip had their various uses. Poultices, both of vegetable and animal nature, were applied. Rest and fasting were naturally practiced. The Indian doctor or medicine man was called a devil chaser for all disease was charged to the wiles of an evil spirit. He acted on the principle that the more hideous he looked and the more frightful he acted, the more successful he would be in chasing the devil away. In his grotesque garb he would attract attention unto himself. His dances, noises and movements of objects such as whirling stones and jingling shells and beads were all elements of the psychology that elicited a faith in him and made the ailing one forget his ills. The collassu or sweat bath, was one of their most effective and ingenius practices. A kind of oven was built up by means of stones and clay large enough to hold two or three persons. There was an opening to admit the person treated and an outlet for the smoke. A fire was built on a bed of stones and kept raging until the stones were very hot. Then the embers were brushed out and the patients would crawl in. The door and smoke flue were closed with skins and water was poured over the hot stones. These sweat ovens were always located near a stream and when the occupants were cooked to the limit of their endurance they would burst out of the enclosure and run to the stream, each one crying at the top of his voice in imitation of the animal of his paahra or totem. When they reached the stream they would plunge in and swim about like rats. After this the patient was immediately wrapped in skins and placed near the fire in the wickwam.

Sometimes marvelous cures were produced by Indian doctors. My grandfather, who lived in Hunterdon County, New Jersey, was afflicted with a cancer of the face. After extended treatment from the family physician without avail, he went to a pow-wow doctor at Reigelsville, Pennsylvania. The old Indian pow-wowed over him and then gave him a dark powder which he was to apply regularly to the infection. This he did and to his astonishment the growth was drawn out of the cheek until it hung down some two inches or more and finally dropped out. The gray-brown dust given him looked like the powder of the puff ball fungus. Whatever it was the treatment was most effective for he lived for many years and there was never any return of the malady.

When death came it caused great consternation in the camp. As soon as a person had died the fact was made known by a howling lament that could be heard for a long distance. The body was bound with hand in front of the face and knees and feet drawn up into a sitting position. The remains were interred in a recumbent position with the head to the south, facing the east. Sometimes they were placed in an erect sitting position decorated profusely with wampum with a pipe in the mouth. When interment was made there was much crying, especially by the women. Some would sing and others sit in silence. Scaffolds were often built over the grave on which food was placed from time to time. Where it was generally known that much wampum was buried with the dead, diligent vigil was necessary to protect the body from thieves.

Wigwams

The Unalachtigos built their houses or wickwams in groups on sandy knolls. They were made by setting two rows of sapling in the ground about ten feet apart and twenty feet long. These saplings were tied together making a vaulted roof. The sides and roof of the house were braced with cross-bars and covered with chestnut bark. A hole was left in the top as a smoke flue. In each end was an opening for a door covered with a flap of skin. There were low platforms within on either side on which the occupants sat or slept. In the center of the floor was a pit for the fire. Split mats of white oak or poplar were used both on the floor and on the platforms. Around the outside of the wigwam a shallow trench was dug for drainage. The village was usually located on the bank of a stream or about a spring. It was often made up of a compact group of houses and surrounded by a palisade of logs. Thus, in case of an attack the occupants were not only fortified but were also provided with a supply of water.

33

Indian Nomads Walked Single File

The Indian by nature was nomadic, especially in the summer season, traveling from place to place as fish and game, fruit and root crops came into season. On the trail he proceeded in single file with the brave taking the lead and the squaw following some distance behind usually overburdened with camp trappings and utensils. By water the dugout canoe, called mushoon, was the medium of travel. This greatly facilitated transportation and progress for the family was easy over the streams of South Jersey.

Dugout Canoes

A number of these dugouts have been reported in the lower part of the State. At Dennisville the old residents speak of one having been seen in the bed of the creek. A large canoe made of cedar was found by Eugene C. Cole in Cape May County at the divide between Dennis Creek and Cedar Swamp Creek near Seaville. The remains of another of sweet gum was located by the writer in the pond at Dividing Creek. Dr. Heritage, of Glassboro, recovered one from the mud of the mill pond at Fries' Mills and kept it in his yard growing flowers in it for a number of years. It was finally thrown on the wood pile and chopped up for firewood. The Gloucester County Historical Society has one from Mantua Creek. Another was found by a Mr. Walford at the point where Third Road crosses a branch of the Great Egg Harbor River near De Costa. Still another has been reported in a swamp at Newport Neck in Cumberland County. Old residents of Tuckahoe remember in their boyhood seeing one on the bank near the point where the present cement bridge spans the river. Diligent search may bring to light others buried in the mud. This is especially likely at the head of tide water.

Tobacco and Pipes

When the Dutch and Swedes sailed up the Delaware they found the Indians cultivating a plant that was very pleasant for smoking. They called it kschatey. The Minsi said glanican. In the early days it was held as a sacred weed with a prominent place in their ceremonies. Pipes were made of clay. They looked much like cigar holders excepting that the bowls were much larger. These pipes were used by individuals in smoking during their leisure. The calumet was the pipe used in common during ceremonies. It was made of stone with stems, an ell or some three feet in length strong enough to lean on. The bowls were massive in size and were so fastened to the stem that they might be used in defense in dealing a deadly blow over the head.

Trails of the Elk

In writing of Indian trails it is of interest to note that originally these trails were not truly Indian. The earliest paths of the Indians were elk trails. At the time of the discovery the elk was common in the State and the network of trails that verily enmeshed the whole region was the result of the Indian capitalizing the lines of communication which the elks had instinctively established. It could not be said of the elk as it was of Sambo, "I don't know where I am going but I'm on my way." His destination was always truly significant. It may be a swamp to browse in, a spring or stream for refreshments, a gap through which to get over the mountain, a convenient place to ford a stream or a shelter under the rocks. To the Indian the elk was "the quarry of his chase," and in his pursuit he learned the lay of the land, established the bounds of his territory, founded villages and utilized the trails made by that animal as fixed lines of communication.

Paths and Trails of the Indians

A careful study of primitive traffic and communication in New Jersey discloses a veritable network of trails throughout the State. The most prominant villages on the coast in the southern area were located, one at the mouth of the Mullica River and another at the mouth of the Egg Harbor River. There were probably more Indians per square mile in the region of Cooper's Creek than in any other locality along the Delaware excepting the Trenton area. In consequence the trail that connected these settlements on the Delaware and the coast grew to be the great thorofare of the region. That is quite like conditions today but the traffic motive is not at all the same.

The villages on the Cohansie appear to have been of great impor tance, and there is also strong evidence that a group of Indians of considerable import lived at the Cape where shell fish were plentiful. In the main, evidence gathered from locations of village sites, characters of implements as to types and material, the nature of terrain and manuscript evidence shows that a trail ran from Camden to Cape May by way of Bridgeton, touching the limits of tidewater on many of the principal streams. There is also indisputable evidence of a path of great importance leading from Beasley Point to Cape May. Another trail traversed the state in an easterly-westerly directory connecting the Bridgeton area with Leeds Point and Pattacoin village at Somers Point. One also meets with occasional reference to a path known as the Shamong trail that ran over the highlands in the middle of the State from Indian Mills to the extreme south over the Old Cape May Road. This thoroughfare passed by Tarkilns Cabin near Belle Plain.

East of Hammonton a path branched off of the Maeroahkong Trail from Camden by way of De Costa and Weymouth to Jobs Point known as Pattacoin Point. Early surveys give the exact location of these trails in the vicinity of De Costa. Naturally enough, in the Cape there was frequent communication between the Coast and the Bay and because of this, it followed that paths connecting points on each of these shores were established. Dr. Julius Way has located these trails in Cape May County and has labeled them with significant Indian names.

The Legendary Migrations and Kings of the Walum Olum

In the legendary classic of the Lenape, the Walum Olum or Red Score discovered by the French scientist Rafinesque, the story is told of their origin west of the Mississippi, how they conquered the Mengwe and the Alligewi, crossed the Na-Missisippu and finally arrived at the Land of the Rising Sun. They came to the Delaware at Trenton, the Stony Falls, Maskekitong. At this point the Unalachtigo divided, some going into South Jersey and some of them to the Chesapeake Peninsula. For a time the Unalachtigo settled at Chichihoki the tumbling bank south of the falls at Trenton where they were governed for many years by the great sachem, Wolomenap, the Hollow Man. Tamenend was their second great sachem and was followed by leaders of the Unami, all of whom took the name Tamanend or Tammany.

The totem of the Unalachtigo was the wild turkey, called Pallouk in the dialect of the Lenape. It means "He who swallows without chewing." A doggerel stanza written by the late Richard Adams, a Unami Lenape of the Indian Territory, sets forth tersely the legend of the Unalachtigo.

"When the tribe was once in danger a wild turkey gave alarm
And the warriors met the bowmen with the fury of a storm.
To a maiden in a vision did the turkey show their plan
And we call all her descendants to this day the turkey clan."

Indian Names of Trails, Settlements, Tribes, etc.

There are a number of names relating to trails and villages still preserved that are interesting and significant. For instance, the path leading from Birds Island on the Bay in Cumberland County to Millville was called "The Indian Going Over." Another north of Tuckerton was known as "The Beach Plum Trail." In the vicinity of Blue Anchor the path from Camden was referred to as "Long-a-Coming." Where the Shamong Trail crossed the Great Egg Harbor River below Winslow we find "Inskeep's Ford." An Indian site at the confluence of the Great Egg Harbor River and Hospitality Branch is known as "The Penny Pot." This name is derived from penipach ghihil'len, meaning, it falls off.

The Caper Tomb, the name of an Island in Cedar Swamp Creek, may be of Indian origin. Cabin Ridge, at Newport, refers to an Indian settlement. The name Yap-shaw found on the Maurice River is derived from Yapeechen, meaning, along the bank. A trail at Port Norris was called Yok wok and one of the streets is still called the Yokwok Road. Dunkydoo applied to the American Bittern may be the Indian imitative name of this bird. Tuckahoe, sometimes written turkey hoe, is truly Unalachtigo.

A certain type of small split basket made by Ann Roberts or "Injun Ann" was appropriately called by the Indians, the "Gulls Nest," since it was often used in gathering gulls eggs.

Though the Indian of New Jersey has gone west we still have words in our vocabulary and names of places and streams many of them with a euphony most attractive to the ear, that will ever memorialize him with his paleface brother. Persimmon, hickory, chipmunk, squaw, chinquepin, kinnikinick, pone, punk, papoose, tomahawk, sumac and sassafras are words that have been directly transferred into our language.

Tribal Territories

The Unalachtigo subdivision of the Lenape held domain over that part of New Jersey lying south of a line from Camden across to the Mullica River. At Cape May were the Kechemeches of 50 men in 1648 with their principal village at Town Bank. Other villages were located at Nummy Town, Diaz Creek, Crooked Creek near the Court House, Mayville, Rio Grande and Cold Spring. In the upper end of the county lived the Tuckahoes with villages at Beasley's Point, Marshallsville, and Head of River.

The Sewapooses were located on the Maurice River called by the Indians the Wa-hat-que nack. There are village sites dotting both banks of this stream from Millville to the Bay. Prominent among them were groups located at the Zane Farm below Heislerville, at Leesburg, Brickboro, Mauricetown, Buckshutem, and Millville. The Siconeeses lived on the Cohansie in the region called Sepahackingh in and about Fairton and Bridgeton. The Alloweys were found on both Stowe and Alloways Creeks in an area called Asamohackingh.

On Mosacksa (Oldmans Creek) and Forcus (Salem Creek) were the Kagkikani Sakins, the Narraticons on Raccoon Creek, the Manteses on Mantua Creek, the Armewamexes on Timber Creek and the Maeroahkong on the Assamoches (Cooper's Creek) from Camden to Medford. In the middle of the State were the Atsions about Indian Mills and the Yacomenshackings possessing the territory from Glassboro south to Vineland. On the coast were the Manahawkins north of the Mullica and the Absegami south of it.

Master Evelin in writing about New Albion (New Jersey) 1648 refers to two Indian kings with forty men each inhabiting the South Cape. He nominates these Indians as Tirans and Tiascons who lived in what is now Cape May County along the coast.

It should be known that the Minsi, originally a sub-tribe of the Lenape, grew to be practically an independent tribe; but the Unami and the Unalachtigo never exercised anything but the closest affiliations. These two sub-tribes of the Lenape were known as the Delawares from the time that the Lenape nation left New Jersey. Even after the westward migration began the Unalachtigo are referred to as the Chickohocki, the name originally held in New Jersey. In order to protect themselves against the Iroquois, the adders as they called them, a league was formed in 1759 banding together the Chikohocki, the Unami, Munceys, Mohicans, and Wappings. (Proud Vol. II, p. 297) From this time forth the identity of the Unalachtigo and the Unamis as separate sub-tribes is lost in the term Delaware, and the tracing of either of these sub-tribes, especially the Unalachtigo, is frought with difficulty.

What became of the Unalachtigos has presented an interesting problem. They are found always closely connected with the Nanticokes. In 1748 with the exodus of the Nanticokes from the Eastern shore many Unalachtigos of South Jersey joined them on their way to the Upper Susquehanna at Chemung and Chenango. By the conference at Crosswicks, February 20, 1758, the Indians from Long Island and south of the Raritan to Cape May were gathered up and placed on a reservation at Edge-Pillock (Indian Mills), Burlington County. There were many Unalachtigos in this group. The reservation was called Brotherton because there were so many different tribes represented: Munsees, Nanticokes, Unalachtigos, Unamis, Raritans, Narragansetts, Montauks, Mohegans, Pequots, Nehantics, Conoys, Tutelos, Saponeys, Shinacocks. Prior to this some of the Unalachtigos had removed to Orange County, New York, near Westbrookville; others were located at Cochecton on the Delaware River and still others on the Juniata in Pennsylvania. In 1802 the Unalachtigos as a part of the Brothertons at Indians Mills went "to eat out of the same dish" with the Oneidas on Oneida Lake.

Notable Chiefs

Although the Unalachtigos were always looked upon as a sub-tribe of the Lenape nation, yet many great men grew out of their gens. Sassoonan the Great, sometimes called Allumapees, was an Unalachtigo of note and a close friend of William Penn. King Shingas during colonial days was the terror of the settlers in Pennsylvania. It is said his deeds of massacre and rapine, if known, would make a blood-curdling tale. Probably the greatest of the Unalachtigo was White Eyes, who lived on the Muskingum in Ohio. In spite of the affiliations of Captain Pipe, chief of the Munsees with the British, Captain White Eyes was sympathetic with the Colonists. He conceived the idea that the Delawares should make up a fourteenth State and join in the cause of the thirteen colonies. It is generally accepted that White Eyes died of smallpox November 10, 1778; however, some say he was treacherously murdered by a Virginia militiaman. His loyalty to the American cause in the Revolution demanded the utmost courage and he will ever be remembered as one of our great patriots.

A notable among the Unalachtigos was Job Chilloway, an Egg Harbor Indian, who took up his abode at Friedenshutten, huts of peace, at Wyalusing on the Susquehanna. He was a scholarly man and served for a number of years as an interpreter in conferences between the Delawares and the Colonial Government.

The last great sachem of the Turkey tribe of prominence among the Delawares was Tha"hu-too-wee'lent, William Anderson who governed the Nation on the White River in Indiana in 1818.

Friendly to New Jersey Colonists

We find the early inhabitants of our State were a simple minded, kindly, trustful, peaceful people. Their bearing on the settlement of the

State was no small factor in paving the way for civilization. They had located the most fertile areas, cleared the ground, established lines of communication and on the arrival of the white men taught the newcomers how to cope with a wild country. They provided food, promoted trade in furs, especially the beaver, showed the colonists how to cultivate corn and tobacco, and gave them ultimately their land as a heritage to an English speaking people.

They were a most unadaptable people and never even to the last accepted the white man's ways. Their fate, as they were pushed westward, was a most inevitable one. They were always facing a dreadful enemy. Those who sympathized with the British were the enemies of the French and the colonists, and likewise those who espoused the patriots' cause were subject to the gunfire of the British. They could do nothing less than suffer martyrdom in the path of progress.

"The warriors race is fading away;
The day of their power and glory is past;
They are scathed like a grove where the lightnings play
They are scattered like leaves by the tempest blast.
They must perish from earth with the deeds they have done;
Already the pall of oblivion descends,
Enshrouding the tribes from our view one by one,
And time o'er the straggling remnants bends
And sweeps them away with a hurried pace,
Still sounding the knell of the warrior race."

—Seba Smith

Gloucester and Salem County Specimens

Plates numbered from A to F—with the exception of F—made from photographs taken by Cochran's Studios, Woodbury, illustrate specimens found in Gloucester and Salem Counties, with a very few exceptions. The most of them now repose in Woodbury. All of the Historical Societies in the Southern part of New Jersey have worth while collections, especially the Cape May County Historical and Genealogical Society at Cape May Court House. Dr. Julius Way and his efficient co-worker Edward Post, have made that Society "blossom like a rose." Its museum is visited by thousands of persons every year. Miss Sibyl T. Jones, the Curator of the Gloucester County Historical Society, states there is more interest indicated in our collection of Indian weapons and implements than all other things we have, together. This statement is the sole inspiration for the "Indians of Southern New Jersey." The writer started work on it considerably over a year ago.

Plate "A"

The implements in plate A are most interesting specimens, viz:

A canoe anchor, found at Mantua Creek, Gloucester County. Weight 17 pounds, groove 20 inches in circumference; at right angle 21 inches. Almost round. A duplex mortar (bowl on both sides). A hoe 7½ inches long and 4½ inches wide at edge. Adze, two pestles, a mall similar to a net sinker, a banner stone with core showing use of hollow reed, grit sand and water to make it; a rubbing stone and a moccasin last. The anchor and the hoe were given to the Gloucester County Historical Society by Hugh Mehorter, of Woodbury. The other specimens are in the collection of the writer. The banner stone was raked up by an oysterman in Morris River Cove. The moccasin last, the rubbing stone, and the small roller stone resting on the mortar were three of the six pieces found at Barnsboro, Gloucester County.

Cochran Studios, Woodbury, N. J.

Plate "B"

The specimens in plate B are a stone pipe of large size, two pottery pipes, two pottery pipes broken, curved adze, celts, knives, scraper, spear heads, gorgets, plummet and a pottery fragment showing the hind quarters of a running animal, like a dog. The soapstone gorget and the plummet were found with four other pieces by a farmer while digging a post hole, near Barnsboro, Gloucester County. All of the specimens were found in either Gloucester or Salem County. They belong to the writer. The pottery pipe at the top left was found near the incised piece of pottery on the farm of the writer's uncle, the late Nathan Steward, on Salem Creek.

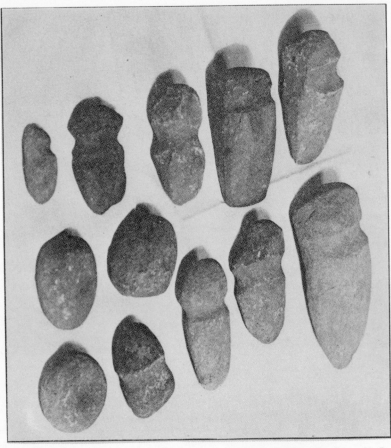

Cochran Studios, Woodbury, N. J.

Plate "C"

The pieces in plate C are representative of different types of axes and hammer-stones of Gloucester and Salem Counties, taken from the collection of the author. The largest ax in the lower right-hand corner is ten and one-half inches long and five inches wide below the groove. The dark colored ax, the second to the right of the small one in the upper left-hand corner, is a very rare double-groove type.

Cochran Studios, Woodbury, N. J.

Plate "D"

The specimens in plate D are mainly from Gloucester County, in the collection of Lewis Lupton. The odd arrowhead in the lower center was found by the writer on his father's farm on Salem Creek, in Mannington Township, Salem County. The pottery fragments are of the common variety of Southern New Jersey.

Cochran Studios, Woodbury, N. J.

Plate "E"

The arrowheads, drills, spears, scrapers and knives in plate E are from Salem County in the collection of the writer. They represent the choice pieces of several thousand specimens.

Plate "F"

The unique specimen illustrated herewith may be a quadruplex mortar, although it has been said to be a sweat stone. The half-tones show both the "obverse" and "reverse" sides.

It was found by George B. MacAltioner, of Woodstown, Salem, Co., in his garden beside a small run. The photograph does not do justice to the object. It seems, like the canoe anchor, so named by the writer, to be the only one yet found. This reminds the writer that Hugh Mehorter presented the Gloucester County Historical Society with a stone hand maul, found in a spring on a farm at Mantua Creek. It is of fine workmanship with a round, flat base about three inches in diameter. It has a cross-bar handle with extended ears to fit the closed hand and flares, bell-like, down to the base. Nothing like it has, so far as is known to the writer, ever been found in the United States. It must have come from some far-away place, possibly the Tahita Islands, but how it ever got lost, in a spring in Gloucester County, is a great mystery.

Miss Dorothy E. Middleton

The three baskets on the top shelf, the one on the second shelf, and the large on Miss Middleton's left-hand are all Papago Indian baskets. She is holding a mission basket, resting on a Sioux Indian drum or tom-tom. At the right of these are some Hopi and Navajo rugs. On her right side are Papago, Shoshone and mission baskets and Klamath Squaw caps.

Miss Middleton is wearing a Blackfoot chief's ceremonial coat, made of white doe-skin and trimmed with beads—or better known as North American ermine—and human hair.

The picture was taken at Trenton at the private showing of the opening of the loan exhibits of the Arts and Crafts of the American Indian. This exhibition was held at the State Museum in December, 1931.

Introduction of the Middleton Collection

Last winter the writer made an address on the "Indian Land Deeds of Southern New Jersey," before the newly organized Archaeological Society of New Jersey, at Trenton. After the meeting he met Miss Dorothy E. Middleton, of Moorestown, N. J., who proved to be an enthusiastic collector of Indian materials. It is due to her generosity that many half-tone plates were made possible for this publication. The photographs of plates 3 to 9, inclusive, were made by the writer. It has been stated that she has the fourth largest collection of ancient Indian relics, in private possession, in the United States.

Miss Middleton has just completed a building to house her noted collection. One of the remarkable things concerning her work is the fact that she has accomplished so much within a short period of time. This is due, in the opinion of the writer, to the fact that her home happens to be located in the midst of an ancient Indian culture, of great magnitude, near the historic Indian Council Fires. The Indians naturally congregated where fish and game were plentiful and the soil was fertile. A glance at the map of New Jersey, or a knowledge of it, will tend to show why Miss Middleton has been so successful in her chosen field of research.

The writer is sure the reader will be greatly interested in the half-tones of a few pieces of her collection and a description of her Indian relics and her method of displaying them.

Plates numbered two to nine, inclusive, show a few of the specimens in the Middleton collection.

My Collection of Indian Relics

By Dorothy E. Middleton

A dream of four years has come true in the form of the new wing to my Ranch House Museum, for the housing of relics of the American Indian, which is a large, sensible, light and well-ventilated hall. It has a gallery around three sides with a stairway on both ends. The walls and ceiling are of beautifully grained fir boards left in the natural color.

The window and door trims and all other woodwork are of dark brown.

There are thirteen cases downstairs against the walls under the gallery. These are filled, with the exception of maybe a dozen pieces, of all New Jersey Stone Relics. On the South end of the room is a replica, half-size, of a Pueblo Oven, and around the room on some of the lower cases are fine examples of baskets made by the Apache, Pima, Mission,

Shoshone, Klickitat and Tlingit tribes, and pottery of the Zuni and Mochi.

On the floor, which is brown in color, are rare Navajo rugs. Under one stairway, near a door which leads into a garden with pools and rockeries, is a table on which are kept some fine editions of books on Stone Age and North American Indians. Here also is a book in which guests are requested to sign their name. In it now are many names from many lands, besides a great many of our States; a record to become more valued and cherished as the years pass by.

Up on the gallery which, of course, has a railing all around and from which hang more rare Navajo rugs, three of which hung in the Boston Museum many years, before their owner sent them to me, are cases the entire length of the gallery, built into the wall, with sliding doors. In some of these cases are more stone relics from New Jersey and also other States as Pennsylvania, Ohio, Virginia, Missouri, Nebraska, North Carolina, Arizona, Kentucky, Tennessee, etc. In the other cases are fine and unsual examples of the arts and crafts of the American Indian; such as a feathered Pomo basket, Sioux trophy coat, Blackfoot chief's ceremonial coat of white doe-skin, peace pipes of the Hopi, Sioux and Blackfoot tribes, beaded apron and hood of the Penobscots, beaded and woven bags of the Winnebagos, beaded necklaces and belts of the Crow tribe, funny carved wooden dish like a fish and a spoon of the Haida Indians of Alaska, and many, many other articles of many other tribes being represented. Most of these articles are one hundred or more years old.

On the walls going up the stairway are old colored engravings of Indian chiefs, made in Philadelphia in 1832 and painted by C. B. King the same year.

Of the stone relics my collection consists of over fifteen thousand pieces, all in good condition. Among these pieces are a stone plow, an Algonquin bowl six inches high and six inches in diameter dug up at Mt. Laurel, nearly six hundred axes, two hundred and fifty pestles, forty perfect bannerstones, one bird stone, one semi-lunar knife (rare), one soapstone gowl, twenty adzes, and six knobbed adzes (scarce), four gauges, fifty celts, sixty mortars, and many thousands of arrow heads, war points, scrapers, drills, knives, spears, moccasins lasts, malls, bolos, nutcrackers, stone beads, pierced gorgets and ornaments, stone and clay pipes, and plumb-bobs. The mortars and pestles range in size from the small ones used for paints—about two inches—to the large pestles of over seventeen inches. The axes vary in size from the smallest of seven-eighths of an inch to one fourteen inches long and weighing seventeen and one-hlaf pounds. This so far as I can learn is the heaviest ax on record in New Jersey. Also there are six double-grooved axes.

The arrow heads are of all materials and are as small as three-sixteenths and one-quarter of an inch in length. These were for bird shot.

The entire collection consists of relics found in South Jersey in Monmouth, Mercer, Burlington, Ocean, Camden, Gloucester and Salem Counties. The most from Burlington County and a few from the other States mentioned before. Probably not over one thousand pieces are from out of this State.

One of the finest pieces, while not a New Jersey specimen, is a Human Effigy Pipe, made of stone and found in a Mound in Tennessee. It is beautifully made and is in perfect condition. A really rare find.

I started collecting seriously about the First of April, 1928. Previous to that we had found a few pieces on our own place. In fact we are still finding them here.

The materials are mostly of native rock of New Jersey; once in a while a piece is found of some foreign rocks such as an obsidion spear, or hemitite plumb-bobs. I have these specimens in my collection with proof of their being found in New Jersey. The obsidion spear was found near Haines Port in Burlington County, and the two hemitite plumb-bobs were found near Port Elizabeth in Cumberland County.

The New Jersey materials for arrows, spears, etc., consist of argillite, flint, jasper, quartz, quartzite and chert. And the materials range differently for the heavier pieces being made of sandstone, trap-rock, diorite, rhiolite, micalite, slates of various kinds and some soap-stone.

All my articles are kept under glass and in separate collections. That is, if I receive one person's small collection of ten pieces, it is tabulated and given a separate shelf or case just the same as if it was a collection of one thousand pieces. In other words, all pieces are kept and marked in their own places so that when a donor comes he may see his lot intact.

Plate Three

At the top is an Algonquin bowl found at Mt. Laurel, Burlington County. It is six inches high and six inches wide, with two small handles.

50

This bowl is mentioned in State Bulletin No. 9, N. J. Archaeological Survey of 1913.

On the right is a soap-stone bowl plowed up near Indian Mills, Burlington County. It has a hole in the bottom, is eleven inches long and seven and one-quarter inches wide. Weight, six and one-quarter pounds.

At the bottom is a very unusual mortar and pestle plowed out together, near Indian Mills. Mortar is beautifully cut out and its bowl conforms to the shape of the natural stone—triangular. It is twelve inches long and nine and one-half inches wide., weighing fourteen pounds. The pestle is of the stubby type and shaped to fit the hand. It is seven inches long and weighs two pounds.

On the left is a rare stone plow, supposed to be one of four known specimens in the United States. It was found near Moorestown. It has a groove all around the top part: was probably pulled by a thong fastened around groove and guided by a forked stick. It weighs six pounds, is eight and one-half inches long and six and one-quarter inches at the widest part.

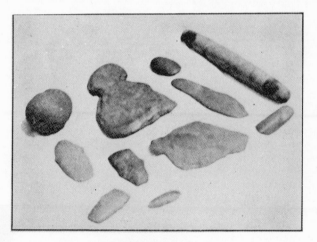

Plate Four

At top left: stone mall or hammer, flat bottom and tapered round top, with a groove completely around the middle, except one inch. Weight, four and three-quarter pounds; height, five and one-quarter inches. Circumference just below groove thirteen and one-quarter inches. Unusual shaped ax. Beveled edges all the way round on both sides. Length from poll to bit thirteen inches, width at center of poll four and three-quarter inches, width at bit eight inches, width at groove three and one-quarter inches, weight six pounds. Unusual specimen.

Small celt from near Mt. Laurel.

Moccasin last from Camden County.

Pestle seventeen and one-quarter inches long found near Marlton, Burlington County.

Small, unsual adze or skinner found on Miss Middleton's farm, Mt. Laurel Road, Moorestown, in Burlington County.

Spade from near Moorestown.

Small green diorite celt, from bank of Cooper Creek, Camden County.

Three, knobbed adzes from Burlington County.

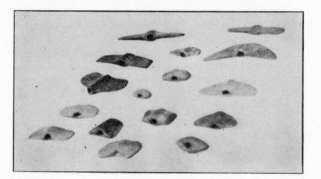

Plate Five

This plate of banner stones, from Miss Middleton's collection, represents a great many types.

Upper right, made of micalite, no hole, from Mercer County.

Next below, Burlington County, drooped wings, drilled through, width seven and one-half inches.

The next three are all from Burlington County and drilled through.

Upper left, Burlington County, has a hole drilled half-way through and is finished complete. Length seven and one-half inches.

Next below, Gloucester County, drilled through.

Next below, Burlington County, drilled through.

Next below, Burlington County, drilled one-quarter through, showing core made by reed used for drilling.

Lower left corner, beautifully polished, length four and one-half inches, tip to tip; drilled through, Cooper Creek. The next two to the right are also from Cooper Creek. Both of them are flat on one side and convex on the other. One of them is drilled through the other just started.

Top center, below two each at right and left top. Made of bluish slate, drilled through, Burlington County.

Next below, drilled half through, finished piece, Burlington County.

The smallest one, drilled through, Mercer County.

The one below, a little to right, seems to be a reworked specimen, drilled through, Mercer County.

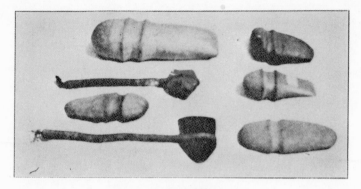

Plate Six

Upper left top. ˙Miss Middleton thinks this is the largest ax of record. It weighs seventeen and one-half pounds, fourteen inches long, and six inches wide at bottom of groove. It is highly polished grey stone and rated as a ceremonial ax; beautifully made specimen, found in a grave at Morgan, Monmouth County, in 1908.

Upper right top. A double groove ax, Mercer County, length seven ·inches, weight five and one-half pounds.

Next below. One of the finest specimens ever found in New Jersey. Found in a grave at Keensburg, Monmouth County, with the bird stone, plate 8, which it matches. Beautifully polished, color light brown, weight three pounds, length seven and one-half inches, three-quarter groove.

Lower right. Made of trap-rock, exquisitely worked, perfect lines, a true work of art; weight three and one-quarter pounds, length eight and one-quarter inches. Found on Miss Middleton's farm, "The Ranch," Moorestown, N. J.

The two tomahawks have modern handles put on them, about forty years ago, to show the method of hafting. The lower one is of black slate and was found near Moorestown, and hafted by the late Harry Chambers. The upper one is a rough ax found in Burlington County.

The ax between the two hafted axes is a peculiar fluted ax, found near Arney's Mount, Burlington County. Has a complete groove. Poll has ridges cut in for decoration and is fluted all the way round. The only one Miss Middleton has ever seen. Weight three and one-half pounds, length eight and one-quarter inches. Very rare and unusual specimen.

Plate Seven

The large object is a dish-shaped mortar, of odd type. It is eight inches long, six and one-half inches wide and weighs six pounds. The piece nearest to it is a three-quarter groove ax with wide, wedge sides, typical of Central New Jersey. It is four and one-half inches long and two inches wide at wedge of poll; weight one pound.

Next, flaked adze of argillite, six and five-eighths inches long.

Above it is a small celt two inches long.

Next, brownstone pipe.

Next, banner stone, drilling started. The last four pieces mentioned were all found in Burlington County.

Next, a rare brown flint drill of the "Stag Horn" type.

Next, a flint arrow of the "Folsom" type, with a groove down the center of both sides. Two and one-eighth inches long.

Next, to right, a fine flint drill, two and one-quarter inches long.

Next, below, an ax made of trap-rock. Complete groove purposely made with curved sides. A curious specimen. Length seven inches, weight one and one-quarter pounds.

Next, small ax with extra wide groove, from Mercer County.

Next, blue-gray quartz spear—same as in plate 8—length seven inches.

Next, lower left corner, knobbed adze.

Next, above, spear of brown flint; three and one-quarter inches long.

The very dark object in the center is a finely polished celt, six and one-quarter inches long.

The last four specimens mentioned are all Burlington County pieces.

Plate Eight

At top left, an arrow shaft polisher, Burlington County.

The mounted card specimens are arrows, spears, scrapers, drills, war points, fish hooks, shell wampum and tiny bird points. The largest piece is an obsidian spear five and one-eighth inches long. All found in Burlington County.

Right of card. A prehistoric stone pipe, in perfect condition, Burlington County.

Below, a plumb-bob of trap-rock, Burlington County.

Next, a skinning knife from Salem County, eight and one-quarter inches long and one inch wide.

Next, a small but finely made mortar.

Next, to left, a lovely spear of blue-grey flint, five and one-eighth inches long, two and one-eighth inches wides at shoulder, Burlington County.

Above, specimen same as Plate 7.

Left, narrow polished brown flint spear, one of the finest known; four and five-eighths inches long, one and one-quarter inch wide at shoulder.

Left, brown flint spear, seven and five-eighths inches long, three inches wide. The one above the narrow spear is made of argillite and is six and three-eighths inches long.

All the spears are from Burlington County. The four small pieces, from left to right, are a perforated stone pendant; a bird stone of finish and material to match ax shown in Plate Six found together in a grave; a magnificent specimen of greatest rarity; a soap-stone effigy pipe, figure of flying goose, similar to Delaware River bridge signs; found in Burling-

55

ton County. The last specimen is almost unique. It is a semi-lunar or squaw knife. Very thin blade, four and one-half inches long, two inches high in middle. Mercer County find. A much larger one is on exhibition in the State Museum, Trenton, N. J.

Plate Nine

Flint and quartz specimens found in the vicinity of Moorestown.

An Early Archeologist

Pierre Eugene Du Simitiere, noted Naturalist, Scientist, Artist, etc., who died in Philadelphia, October 8th, 1784, was a collector of Indian relics. His writings show that he wrote to Isaac Zane, of the Marlboro Iron Works, Williamsburg, Va., about a stone relic found at Egg Harbor, N. J.

In November, 1779, he made mention of a hideous painted face mask of wood, found with about forty more, all different, at the Indian town of Chemung, of the Six Nations, destroyed by General Sullivan. He termed them Manitoe faces. He recorded a celt found on the farm of Joseph Cooper, opposite Philadelphia; A Shawnee Indian vocabulary furnished to him by Col. Richard Butler; a decorated Indian scalp sent to him by the President and the Supreme Executive Council of Pennsylvania in 1782.

In New Jersey, at the Delaware Falls, near Trenton, where the bank of an Indian graveyard had washed down, he found a variety of Indian utensils. This place was opposite an Island just above the mouth of the creek "that Trenton mill is built on."

Near the above place on the property of Col. John Cox, formerly belonging to Dr. William Briant and Robert L. Hooper, he mentions hatchets, chisels, pestles, arrow heads, etc., as being plowed up in abundance in 1777-80.

On the plantation of Dr. John Kearsley, about four miles from Philadelphia, he obtained the largest stone hatchet he had ever seen.

He was given by General Sullivan, an Indian face of red stone of the same character as Indian Chiefs' pipes. It is evident that the Indians made common use of these masks. This particular one was suspended around the neck of the King of Kanadasego, killed in action August 29, 1779.

Du Simitiere also had a broken stone, shaped like a shoemaker's last, supposed to be of Indian workmanship. It was found in a meadow near the Falls of Schuylkill. This reminds me that the first one I had ever seen was found with five other pieces of stone by a man digging a post hole near Barnsboro, N. J. The Bureau of Ethnology at the time wrote there was no such thing hereabouts as a moccasin last, but finally said such might be possible because one of their men had found them in the southwestern part of the country. I have seen six or more of them since.

Du Simitiere was a poor man but one of tremendous energy and achievement. In almost every field of research one will find his never-to-be-effaced tracks.

A sketch of him and a glimpse of his efforts are to be found in the 1889 Magazine of History.

Indian Marks or Signatures

Indian marks or signatures of the early kings and chiefs are rarely met with at present. Some few original Indian land deeds, with signatures, have escaped destruction and all of them found by the writer, with one exception, in the hands of a county historical society are reproduced herewith.

Necomis, of Salem County, seems to have used the moon for his mark. This was one hundred and fifty years before the Indian woman of that name was mentioned in Longfellow's Hiawatha as the daughter of the moon.

It will be noticed that the early Indians did not always use the same mark and it is probable that the most of them had no standardized signature.

The spelling of the Indian names at the hands of the purchasers was just as irregular as the marks of the Indians.

The writer thinks that the two half-tones of thirty-three Indian marks, one of the most interesting features of this publication. If any reader can send him tracings of other marks of Indians for future use they will be highly appreciated. By reference to the section pertaining to the deeds the reader may locate the present location of the deeds signed by the Indians with their marks. Mahamecun was the alleged cousin of King Charles, and not the King himself, as stated on the plate of signatures.

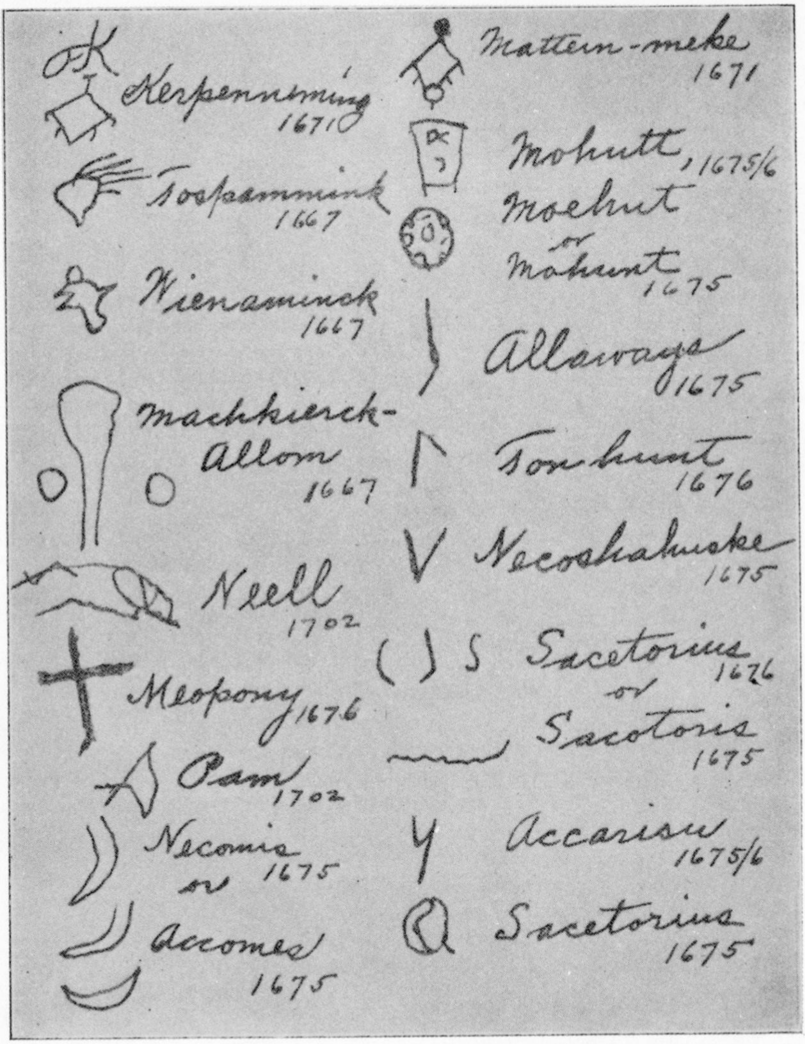

Kerpennening
1671

Sospammink
1667

Wienaminck
1667

machkierck-
Allom
1667

Neell
1702

Meopony
1676

Pam
1702

Neconis
or
1675

Accomes
1675

Mattern-meke
1671

Mohutt, 1675/6
moehut
or
Mohunt
1675

Allaways
1675

Ton hunt
1676

Necoshahuske
1675

Sacetorius
or
1676

Sacotoris
1675

Accarisu
1675/6

Sacetorius
1675

Toospeesmick
1665

Wennamink
1665

Seketarius
1683

Whiscatne
1702

Meopeny
1675

Leneaquah
1697

Soswoaton
1697

Quetayaha
1697

Mahamecun
alias   1697
King Charles

Chochanaham
1675/6

Opa Hoge
1701

Hokgo Sooway
1701

Hicmorn
1701

Indian Deed for Land Between Assinpink and Rancocus Creeks (abstract)

One Conveyance or Deed bearing date the 10th of October 1677 made by Ahtahkones, Nanhoosing, Okanickhon, Weskeakitt, Pecheatus, Kekroppamant, Indian Sackamackers of the one pte to Joseph Hemsley, Robert Stacy, William Emley, Thomas Folke, Thomas Ollive, Daniell Wills, John Pennford & Benjamin Scott of the other pte, that Tract of Land lying along the River Dellaware from & between the midstream. of Rankokus Creeke to the southward & the midstream of Sent Pinck Creeke at the ffalls to the northward, and bounded to the eastward by a Right Lyne & extending along the Countrey from the uppermost head of Rankokus Creeke to the Partition Lyne of Sr. George Carteretts Right against the uppermost head of St Pinck Creeke for the consideracons of Forty six fadome of Duffelds, Thirty blanketts, one hundred & fifty pound of powder, Thirty Gunns, Thirty kettles & Thirty kettles more instead of Wampam, Thirty Axes, Thirty Howes, Thirty aules, Thirty needles, Thirty looking glasses, Thirty pair of Stockings, Seaven anchors of Brandy or Rum, Thirty knives, Thirty barres of lead, Thirty six rings, Thirty Jews harps, Thirty combs, Thirty braseletts, Thirty bells, Thirty tobacco toungs or Steeles, Thirty pair of Sissrs, Twelve tobacco boxes, Thirty flints, Tenn spoonfulls of red paint, one hundred of fish hooks, one gross of tobacco pipes & thirty shirts to them paid.

Witnessed & executed before Henricus Jacobs, Marmaduke Rosendale, Mathew Smith, Alex Watts, Tickahoppenick, Attarrumhah.

Vol B part 1 of deeds, page 4.

Indian Deed for Land Between Rancocas and
Big Timber Creeks (abstract)

One Conveyance or Deed bearing date the 10th of September 1677 made by Katamas, Sokappie, Peanto alias Enequeto, Rennowighwan, Jackickon, Indian Sackamackers of the one pte to Thomas Ollive, Daniell Wills, John Pennford, Benjamin Scott, Joseph Hemsley, Robert Stacy, William Emley, & Thomas Folke of the other pte of that Tract of Land lying along the River Dellaware from & between the midstream of Rankokus Creeke northward, and the midstream of Timber Creeke southward & bounded to the eastward by a right lyne & extended along the Country from the uttermost head of Rankokus Creeke to the uttermost head of Timber Creeke for the consideracon of fforty six ffordome of Duffelds, Thirty blankitts, one hundred & fifty pounds of powder, Thirty gunns, Two hundred ffadomes of Wampam, Thirty Kettles, Thirty Axes, Thirty small howes, Thirty aules, Thirty needles, Thirty looking glasses, Thirty pair of Stockings, Seaven Anchors of brandy, Thirty knives, Thirty barrels of lead, Thirty six rings, Thirty Jews harps, Thirty combs, Thirty Braseletts, Thirty Bells, Thirty tobacco toungs, Thirty pair of Sissurs, Twelve Tobacco boxes, Thirty flints, Tenn pewter spoonfulls of Paint, one hundred of fish hooks & one gross of pipes to them paid.

Witnesses & executed before Thomas Watson, Andrew Swanson, Swan
Swanson, Lasey Swanson.

Vol B part 1 of deeds page 4.

Indian Deed for Land Between Big Timber and Oldmans Creeks
(abstract)

One Deed bearing date the 27th of September 1677 made by Mohocksey, Tatamerkho, Apperinyes, Indians of the one pte To John Kinsey,
Thomas Ollive, Daniell Wills, John Pennford, Benjamin Scott, Joseph
Hemsley, Robert Stacy, William Emley & Thomas ffolke of the other
Pte of that Tract of Land from & betweene the midstreame of Oldmans
Creek to the southward & the midstreame of Timber Creeke to the
northward and bounded to the Eastward by a Right Lyne extended along
the Countrey from the uppermost head of Oldmans Creeke to the upper-
most head of Timber Creeke for the Consideracon of Thirty Matchcoats,
Twenty Gunns, Thirty Keettles & one great one Thirty paire of Hose,
Twenty ffadome of Duffells, Thirty Petticoates, Thirty Indian Axes,
Thirty narrow Howes, Thirty Barres of Lead, ffifteene small Barrells
of powder, Seaventy knives, Sixty pair of Tobacco Toungs, Sixty Sissers,
Sixty Tinshaw looking glasses, Seaventy Combs, one Hundred & Twenty
aul blades, one Hundred & Twenty ffishhooks, Two grasps of Red paint,
one hundred & Twenty needles, Sixty Tobacco boxes, one hundred &
Twenty pipes, Two Hundred Bells, one Hundred Jewes Harps, Six
Anchors of Rum to them paid.

Witnessed & executed before Robert Wade, James Saunderland
James Yesteven, Samll Lovett & Henry Reynolds.

Vol B part 1 of deeds page 3.

Indian Deed for Fenwick's Lands Between Oldmans and Salem Creek

Be it known unto all people and persons whatsoever by these presents that we Tospaminke and Wenaminke the true and undoubted owners
(as by Natural right and interest of all and singular that tract of land
scituate, lying and being within the Province of New Cesaria or New
Jersey and butteth and boundeth as is herein expressed viz from the
mouth of the Creeke heretofore known by the name of Forcus Creeke
or Game Creek and now called and knowne by the name of Fenwick's
Creeke or River and so it runeth up Delaware River unto a certaine
Creeke on the Easterly shore called and knowne by the name of Mas-
acksa running up the said Creeke unto the head thereof unto the Mayne
land and so all along to the head of the north and West Branch of Fenwicks Creeke or River and so down the North West side of the said
River or Creeke unto the mouth thereof as aforesaid.

Send greeting Wherefor Wee the said Tospamynk and Wenamynke
have received of John Fenwick Esq., one of the Lords or Chief proprietors of the said Province of New Cesaria two ankers of Rum, eight

Knifes three payer of cissors and divers other English Commodities as appear by a particular of the parcels under the hand of the said John Fenwick for and in consideration of the said tract of land the receipt whereof we doe hereby acknowledge and accordingly doe acquitt and discharge him the said John Fenwick his heirs, Executors administrators and assignes thereof and of every pte parcell thereof by these presents And for and in the consideration thereof and for divers other good and valuable causes and considerations we thereunto moveing have given, granted remysed released and quitt claimed and by these presents do for us and our heirs, give grant remise release and forever quit claim unto him the said John Fenwick (in the full peaceable and quiet possession and seisen being) and to his heirs and assignes forever all the estate right title interest use claim & demand whatsoever which we The said Tospamynke and Wenamynke now have or had or which our heirs, Executors and administrators or any of our Natives att any tyme hereafter shall or may have or claim of & to all the above fore mentioned tract of land or of and unto all and every or any pte or parcel thereof by force and virtue of any act done by us or either of us to the contrary And we the said Tospamynke and Wenamynke do covenant and grant for us and either of us for our heirs and people to and with him the said John Fenwicke his heirs administrators and assigns that he and they shall have hold and enjoy the said tract of land together all the said creeks, marshes rivolets woods trees, mynes, mineralls and hereditaments whatsoever or thereunto belonging peaceably and quietly without the least lett hindrance molestation or troubles whatsoever of us our heirs or people we hereby covenansing and promising to and with him the said John Fenwick his heirs and assigns that we shall and will from time to time well and sufficiently warrant them and defend them forever.

In Witness Whereof we the said Tospamynke and Wenamynke have this second day of the first month commonly called March sett our hands and seales in the year 1676 by the new style

Signed, sealed
and delivered in the
presence of   Tospamynke   by his mark
(torn)   Wenamynke   by his mark
rth   X   Johnson
rtt   X   Jecocke

Copied from a photostat furnished by the New Jersey Historical Society.

John Fenwick's First Purchase November 17, 1675

Be it known unto all persons & people whatsoever that Wee Mehawcksy, Allowayes, Myoppeney, Saccatorey, Necomis and his mother Necosshesco and Mohutt the undoubted owners of all Lands hygh & lowe, Meadows, Rivers, Creeks, Lakes, Brooks, Timbers & whatsoever else doe thereunto appertain within the circute & bounds of all & every pte &

62

pcell of the sd Land called or known by these names as followeth which lyeth by the Creek or River called or known by the name of Game or Forcus Creek to the uttermost extent thereof up into the Land or Continent and from the mouth thereof which runns into the River called & known by Dellaware River downwards to a certain Creeke or River commonly called or known by the name of Canahockonck adjoining to the lands belonging unto Chahanzick and soe running up the said River of Cannahockinck from the mouth thereof up to the furthest extent thereof into the mayne Lands & soe to cross from the head thereof unto the head of the said River or Creeke called as aforesd the Gamye or Forcus Creeke, all which tract of Land hygh & lowe meadows Rivers Creekes Lakes Brooks Timbers and whatsoever else thereunto belong or appertain situate lying & being within the province of New Casaria or New Jersey in America being so called or known by the English Nation and others, Wee do hereby for & in consideration of Wooling, linen apparell and divers other comodities agreed upon, Covenant promise & grant both for ourselves & People our heirs and Successors for all other Indyans whatsoever which may any wayes pretend or claim a right unto any of the Tract of Land soe granted by us as aforesd which wee doe resign up together with our whole right title claim & interest therein unto John Fenwick Esqr Chiefe Proprietor of the one halfe of the Province of New Cesaria or New Jersey aforesd his heirs & assigns forever together with all & every th appertaines & hereditaments thereunto belonging whatsoever to be disposed of & possessed at his or their will & pleasure, excepted always out of this grant the Plantacons in which they now inhabit in for & untill such time only as they shall think fitt to remove from the same. So for most full Confirmation of all the premises Wee have allso according to the English custome Rattifyed and confirmed this our grant & sale as aforesd under our hands & seales this seventeenth day of the ninth Mo. in the year 1675.

Signed, Sealed & delivered	
in the presence of	Soccatorey.
Foppe Out Hout	Necosshesco.
Michael Baron	Mohutt.
Francis Whittwell	Necomis.
Thomas Turner	Allowayes.
	Mohoppeney.
	by their marks.

Salem 1 of Deeds, page 18 (Sec. of State of N. J.)

From a "true copy" by Fenwick of the above deed in the possession of the New Jersey Historical Society. Some of the Indian names are spelled thus: Myhopponey, Shuccatorey, Necosshehesco, Neccomis.

Supplementary Indian Agreement with Fenwick

Whereas Mahocsey, Mohopponey, Allawayes, Necomis, Sacotorys, Mohutt & Necoshahuske sold to John Fenwick Esqr all that Tract of Land mentioned in a deed under their hands & seales for divers pcells of goods mentioned in a skedule under the hand of sd John Fenwick &

63

now in the keeping of them or some of them, and Whereas they have received of him the sd John Fenwick divers & sundry of the sd pcells of Goods & being willing to receive in lieu of all the remainder unpayed, One Intyre peece of Duffill about the quantity of the former piece which was payd them by the sd John Fenwick as allso four Gunns to be payd as followeth Viz Two gunns in hand & two gunns more when Mahoacksey, Mahopponey, Necoshahuske shall seal together with them the sd Necomis, Allowayes, Socatarys & Mohutt, as allso to deliver back the sd note of particulars, all which they the sd Necomis, Secatories & Allowayes whose hands are hereunto subscribed, do promise to deliver together with their discharge & full acquittance for the same & do accordingly acknowledge the receipt of the sd peece of Duffills or Matchcoate together with two gunns for & in full of all the sd goods & doe Ingage to save harmless the sd John Fenwick his heirs & assigns of & from the claim demand trouble & molestation of all psons whatsoever for & concerning any other right title claim that any other hath or can have unto the sd Tract of Land sold as aforesd unto the sd John Fenwick his heirs & assigns forever.

In witness thereof the sd Necomis, Sacatorys & Allowayes have sett their hands this eighth day of the Eleventh Mo. in the year according to the New Style **1675**.

Witness—

Accaroya by mark
Richard Guy
Rennear Vanhist
Thomas Watson.

Sacatorys.
Necashahuske
Necomis.
Mohut.
by their marks.

Salem 1 of Deeds, page 19

The original in possession of the New Jersey Historical Society shows some few slight variations in spelling and words from the above copy from the deed book.

Seketarius Deeded Land to William Penn

At Philadelphia on December 19, 1683, Seketarius and six other Indians gave a deed to Governor William Penn for the land between Upland and Christina creeks in Pennsylvania. The deed was witnessed by Thos. Holme, John Morris and John Songhurst. This is the only Indian I have found that signed deeds for land on both sides of the Delaware river. Mrs. Janvier of New Castle has the deed. I have seen four signatures of Sacatorius and no two were alike although the three signatures of Nicomus I have seen were all similar. Years after I had copied a Sacatorius signature I mistook it for some of my own shorthand notes which I could not decipher and did not find the difference until I re-examined the Indian deed.

Confirmation Indian Deed to Fenwick

Be it known to all psons whatsoever by these Presents that Wee Mohutt, Alloways, Myhopponey, Saccotorey, Necomis, Necosshehesce his mother the true & undoubted owners (as by natural right & interest) of all that tract of Land as it lyeth bounded within the Creeke hereafter named (Viz.) from the mouth of the River or Creek heretofore called by the Natives the Game Creek & by others Forcus Creeke and now by the English itt is to be hereafter called Fenwicks River or Creeke which runeth into Dellaware River & so from the mouth thereof down the easterly shore of the sd Dellaware River to the Creeke which is also called by the Indians Cannahockink Creeke & so runeth up that Creeke into the main land now called by the English the province of New Cesaria or New Jersey in America, to the uttermost extent of the said River or Creeke & of every branch thereof, and from thence to the utmost extent of the Creeke called Allowayes Creeke and of every branch thereof, and from thence to the outmost extent of the sd Creeke called Game Creeke or Forcus Creeke, and all and every the branches thereof by whatsoever name or names the same have heretofore been or is called & known, and so down the sd Game Creeke to the mouth thereof as aforesd, all which tract of Land together with the said three Creeks with all other Rivers, Creeks, Lakes, Brooks, Lands high & low Meadows, Marshes, Timber trees & all other trees & woods & whatsoever else doth thereunto belong & appertains, Wee the sd Mohutt, Allowayes, Myhoppaney, Soccatorey, Necomis and Necosshesco for & in consideration of a good & valuable quantitye of Rum, Tradeing cloth, shirts, hose, shoes, Gunnes, lead, pouder and other English apparrell & Comodities well & truly payd unto us by John Fenwick Esqur Lord or Chiefe Proprietor of the one halfe of the sd province and received by us, hereby give grant Alien bargain & sell by the presents, have given granted aliened bargained & sold unto him the sd John Fenwick his heirs & assigns forever, and wee accordingly resign up our whole right title claim & interest therein & in every pte & pcell thereof together alsoe with all the Dominions & Terrytories thereto belonging, to be disposed of & possessed by him the sd John Fenwick his heirs & assigns at their own wills & pleasures forever, only the Plantacons in which wee now dwell, situate within the sd Tract of Land, excepted for and until such time only as wee shall think fitt to move from the same: We behaving ourselves in the mean time honestly & peaceably towards the English people without any molestation of contriement agst them, or abetting any that shall any wayes molest them or any of them, especially them as shall come to inhabit within the sd Tract of Land by the order & appointment of him the sd John Fenwick his heirs & assigns, and we the sd Mohutt, Allowayes, Myhoppaney, Saccotorey Necomis & Nicosshebesco, and every of us do hereby as aforesd renounce all our right title claim or plea to the sd purchased Lands, Creekes, & all other the premisses before menconed to be hereby sold and all things therein contained, resigne all our right thereunto to him the sd John Fenwick his heirs & assigns forever, and for a token of our making good of all & every the premisses, were and every one of us have voluntarily put the sd John Fenwick in the peaceable possession of the sd Tract of Land, and for the most full confirmecon of all the premises according to a former contract made by us the seventeenth day of the nineth Mo last Wee have according to the English custom Rattifyed & confirmed this our grant & Sayle as aforesd under our hands & Seales this fourteenth day of the first Mo commonly called March in the year One Thousand Six hundred & seventy six by the new Style.

Signed Sealed & delivered
in presence of
William Malster
Marcus Ellers
Richard Wittaker

Mohawoksey X the King	Mohutt	X	(seal)
Thomas White	Allowayes, Myhopponey X		(seal)
Richard Noble	Saccatorey X Necomis		(seal)
William Johnson	Necosshehesco	X	(seal)
Thomas X Watson			

Elizabeth Adams
Occarous X his mark
Opur Mohawkeseys X brother mark
Wittan Awke X
Salem 1, page 42

This original deed is in possession of the New Jersey Historical Society.

Indian Deed to Fenwick for Little and Great Cohanzick

Be itt known unto all People & persons whatsoever by these psents that Wee Mohawksey, Mohutt, Newsego, Chochanaham, Torucho & Shacanum the true & undoubted owners (as natives) of all & singular the Tract of Land called & known by the name or names of Little Chohanzick & Great Chohanzick, sittuate, lying & being within the Province of New Casaria or New Jersey with all & every their appurtenances, and as it is herein butted & bounded (That is to Say) from the mouth of the Creeke called Cannahockincke, down the River of Dellaware unto the Bay to a certain Creeke next unto the mouth of Chohansey Creeke or River, Southward called Weehattquack and thence to the head of the sd Creeke, and from the head or utmost extent thereof in a direct lyne to the head & uttermost extent (into the mayne land) of Chohanzey Creeke or River, & from thence in a straight lyne to the head & utmost extent of the sd Creeke called Cannahockinck, & soe down into the sd River of Dellaware. Now these psents witnesseth, that Wee the sd Mohawksey, Mohutt, Newsego, Chochanaham, Torochoe & Shackanum doe give grant bargain & sell & by these psents have given granted bargained & sold, all & every the sd Tract of Land of Little & Great Chohanzick with the appurtenances soe butted and bounded as aforesd unto John Fenwick Esqr

One of the Absolute Lords Proprietors of the sd Province of New Casaria or New Jersey herein America, his heirs and assigns forever for & in consideration of fouer Anchers of Rumm, twelve Matchcoates, and divers others pcells of Goods well & truly paid & Sattisfied unto us by him the sd John Fenwick the receipt whereof Wee doe hereby acknowledge & accordingly doe aquitt and discharge him the sd John Fenwick his heirs Executors Administrators & assigns thereof & of every pte & pcell thereof by these psents and Wee the sd Mohawksey, Mohutt, Newsego, Chechanoham, Torocho & Shackmun,doe for ourselves Successors & people Covenant promise grant and agree to & with him the sd John Fenwick his heirs Executors Administrators and assigns that he & they and every of them shall from henceforth have hold & peaceably enjoy the sd Tract of Land now called as aforesd Little & Great Chohanzick together with all rivers, creeks, marshes, woods, Timber trees heriditaments and premisses thereunto belonging with all and every the appurtenances whatsoever without the lett trouble Interupcon of us or any of us or of our heirs or Successors or any of them or any other Indyans whatsoever Wee hereby freely and willingly for the consideracons aforesd, Doe hereby accordingly resign up the sd Tract of Land called Little & Great Chohanzey bounded soe as aforesd together with our whole right title clayme & Interest therein unto him the sd John Fenwick To have and to hold the sd premises to him his heirs Exec, Administratrs and assigns forever, and to & for no other use intent or purpose whatsoever, and accordingly to be disposed of and possessed by him the sd John Fenwick, his heirs Executors Administrators & assigns, Excepted always out of this grant the town & Plantacons in which they the sd Indyans now inhabitt and useth for & untill such time only as they shall think fitt to remove from the same.

Soe for the most full confirmation of all the premissess wee have according to the English cutom rattified & confirmed this our grant & Sayle as aforesd under our hands & seales this Sixt day of the Twelveth Mo. 1675/6. (February 6, 1675/6). Year began March 25.

Signed Sealed & delivered
in the presence of us by
Henry Parker
Richard Noble
John Smith
Richard Guy

Chochanaham.
Mohutt.
Torocho.
by their marks

Salem 1 of Deeds, page 20
Original in possession of New Jersey Historical Society.

Abstract of Indian Deed for Land Between Cohansey Creek and Morris River

To all persons to whom these psents shall come, Know yee that wee Mauhauxett, Cuttenahquoh, Keshuwicon, Attahissha, Sucolana & Awhehon Lords & Indian Proprietors of all the Tract of Land lying & being betwixt Cohannsick & Prince Morris River in the Province of West New Jersey for & in consideration of the sume of Twenty Guns, Twenty Blanketts, Twenty Stroudwaters Red & Blew Coats, Thirty Coats of Red & Blew Duffels, Ten Shirts, Seaven pounds Ten Shillings in money, Fifteen small kettles, Five Hats, Twenty pair of Stockings, Ten pair of Shooes, half a barrell of powder, Fifty barrs of lead, Two hundred flints, Twenty five pound of shott, fforty knives, fifteene Hatchetts, Ten Howes, ffive Addses, ffive drawing knives, ffive swords, Two Handsaws, Thirty looking glasses, Thirty Combs, Thirty pair of Tobacco tongues, ffifteene Coats, one hundred fish hooks, Three hundred kneedles, Twenty Tobacco boxes, Thirty Aules, one Barrell of Rum, one Barrell of Beer, Twelve pounds Red Lead, half a hatt full of Beads, one double hand full of Jews Harps, one hundred pipes, one hatt full of nayles, Six pistolls, Thirty pair Sissers to us the aforesd Lord & Indian Proprietors in hand paid by Jeremiah Basse, Agent to the Hon'ble the West Jersey Society for & in the behalf of the sd Society the receipt whereof wee hereby acknowledge all that Tract or pcell of Land lying & being betwixt Cohannsick Creek & Morris River begining at ye head of Cohannsie att the Horse Road & soe running on a straight lyne crose ye neck to a Cabbin comonly known by the name of (left out) Cabbin on Morris River & soe down ye river bounding with ye middle of the river to the mouth of the River & soe along the Bay bounding therewith to Cohannsie River & up the said River bounding with the same to the place before named (excepting a neck of land lieing betwixt Cohansie & the River of Tweeds (?) begining att Willm Johnstons plantation on Cohansie & soe continued by a Line to the plantation of Thomas Shepheards including all the Lands below their Plantations in the said neck.

As Witness our hands & seales this nineth day of June Annoq Dom 1693.

Signed Sealed & delivered
in ye psence of us

George Taylor	Mauhauxsett with a seale
Obadiah Holmes	Cuttenoquoh with a seale
Joseph Holdin	Kesshuwicon with a seale
Rineare Vanhist	Attahissha with a seale
The mark of Apahon	Sucolana with a seale
The mark of Matchuos	Awhehon with a seale
The mark of Youthlon	By their several marks.
The mark of Jamickcoock	
The mark of Swanamemigh.	

Vol. B, part 1, page 325 of Deeds. Sec. of State of N. J.

Abstract of Indian Deed for Land Between Stephens Creek and Little Egg Harbor and Cape May Point Including All of Cape May County

To all People to whom this psent writing shall come, Wee Sakamoy, Tamahocks, Apauko, Squehon, Ossemahaman alias John Monoockomen alias Mr. Tom Mannime, Indian Sachimackers & owners of the Tract or

68

Tracts of Land hereinafter particularly named within the precincts of West Jersey, send Greeting—

Knowe ye that the said Indian Sachimackers & owners of the said Tract or Tracts of Land hereinafter mentioned for and in consideration of Ten script matchcotes, Twelve blew & red matchcoates, twelve strowd water matchcotes, twelve strize cotes, Ten Kettles, Twelve shirts, twelve pair of Stockings, Thirty two knives, Forty five shillings in Silver, Twenty barres of Lead, Ten tobacco boxes, one Runlett of Shott, halfe hundred of powder, fower pound of Red Lead, one Grosse one dozen of pipes, Two Capps, fower adzes, five hand sawes, Two hundred flints, Tenn Gunns, Ten axes, Tenn howes, fower drawing knives, Twelve looking glasses, five Steeles, Eighteene auls, twelve combs, six Jews harps, Sixteen Gallons of Rumm, one barrell of beere, two paire of Shooes & two calicoe neck clothes, By Adlord Bowde now of Burlington in the said Province merchant for & on the behalf of Daniel Coxe Esqr Governor of & Proprietor of ye said Province att & before the sealing & delivery hereof to them ye said Indian Sachimackers & owners well & truly in hand paid whereof & wherewith they doe hereby acknowledge themselves fully contented & satisfyed have granted bargained & sold alyened enfeoffed & confirmed, and by these psents doe fully clearly & absolutely grant bargain & sell alyen enfeoffe & confirm unto the said Adlord Bowde for & to ye only use & behoofe of the said Daniell Coxe & of his heirs & assigns forever, All that or those Tract or Tracts of Land layd forth marked & bounded as hereafter followeth, That is to Say, from the mouth of Stephants Creek on ye north side of Dellaware Bay by a Lyne runing northerly to ye outmost flowing of the Highest flood in Petequlick or nixt river or creek being westerly from Little Egg Harbour, thence down sd Creeke or river of Petaqueick to little Egg harbours most easterly point upon the sea, then southwesterly by the Sea to Cape May and round along sd cape towards Dellaware Bay westerly thence northeasterly along sd bay to ye bottom thereof, Thence westerly to ye mouth of foresd Stephants Creek to ye first begining or departure purchased & paid for by Andrew Robeson on behalf of ye foresd Bowde & Coxe, and allso all & every the mines, mineralls, woods, fishings, hawkings, huntings & fowlings & all & every ye Rivers, Rivoletts, Creeks, Isles, Islands, Lakes, Ponds, Marshes, Swamps & Meadows.

In witness whereof ye sd Indian Sachimackers & owners have hereunto sett their hands & affixed their Seales ye thirtieth day of April Anno Dom 1688 & fourth yeare of ye reigne of James ye Second of England King.

Sealed & delivered in
the psence of us
Andrew Robeson
Caleb Carman
Andrew Robeson Jr.
Henry Jacobs Fulconbridge interpreter

Saquemoy, Tamahack
Apauko, Sweikhon,
Oshemahamon or John Monoockomen or
Mr. Tom Nunimi by their several marks.

69

by their marks.

Vol B part 1, page 202

Mr. Tom Nunimi was undoubtedly the Indian of Cape May now called Nummy. He seems to have had several names.

The Indians, like the Negroes, soon adopted the names of the whites.

Indian Deed for Land Beyond the Headwaters of Creeks Flowing into the Delaware (abstract)

To all People to whom this psent writing shall come, Wee Molhunt, Jakomis, Maundicon, Manickopon, Tapashito, Tessiokon & Sickajo, Indian Sachimackers & owners of ye Tract or Tracts of Land hereinafter particularily named with ye Province of West Jersey send Greeting—

Know ye that ye said Indian Sachimackers & owners of the said Tract or Tracts of Land hereinafter menconed for & in consideration of Eight stript matchcoats, Eight Matchcoates, Seven stroud water matchcoates, eight Frize Coats, Ten Kettles, nine shirts, nine pair Stockings, thirty knives, fourty five shillings silver money, Fifteen barres lead, Ten tobacco boxes, one Runlett Shott, half barrell of powder, four pound red lead, one grosse one dozen of pipes, two Capps, three adzes, three hand sawes, two hundred flints, Ten Guns, eight axes, Seaven howes, three drawing Knives, twelve Glasses, five steeles, eighteen aules, twelve combs, six Jews harps, Ten gallons Rum & Cask & one barrell of Beere, by Adlord Bowde now of Burlington in the Province aforesd, merchant for & on behalf of Daniell Coxe Esqr Governor of & Proprietor in the sd Province att & before the sealing & delivery hereof to them the sd Indian Sachimackers & owners well & truly in hand paid where of & where with they doe hereby acknowledge themselves fully contented & satisfyed, have granted, bargained, & sold alyened enfeoffed & confirmed, and by these psents doe fully clearly & absolutely grant bargain & sell alyen enfeoffe & confirm unto the said Adlord Bowde, for & to ye onely per use & behoofe of the sd Daniell Cox & his heirs & assigns forever, All that and those Tract & Tracts of Land laid forth marked & bounded as hereafter followeth, That is to Say, Begining at the easterly point upon Cohanisck Creek upon the Bay, thence along sd Creeke & soe to Oldmans Creek, thence to Timber Creeke & from thence to the upmost flowing of the Tyde in the River or Creek nixt running down to Little Egg Harbour, thence along the Lyne lately purchased of Sequamoy, Tamahack & others to the Bay of Dellaware by Stephants Isle, thence along sd Bay or Coast up again to the first point upon Cohansick Creek, Excepting out of this Stephants Isle & four thousand acres of land formerly sold near the mouth of Cohansick Creek purchased & paid by Andrew Robeson for ye use & behoofe aforesaid.

In Witness whereof the sd Indian Sachimackers & owners have hereunto sett their hands & affixed their Seals the twenty fourth day of June

70

Anno Dom 1688 & fourth yeare of ye reigne of James ye Second of England or King.

Signed & sealed & delivered
in ye presence of us
Andrew Robeson
Thomas Sharp
Alexander Menzies
Henry Jacobs fulconbridge

Molhunt
Jakomis
Manickopon
Maundiron
Tapashito
Tessiokon
Sickajo
by their several marks

Vol B part 1 of Deeds, page 203

Molhunt is probably the same as Mohutt.

Abstract of Patent for Salem County Land

Indian deed from Tospeesmick and Wennanminck for land along the Delaware River, dated 14 Oct. 1665 to Fopp Jansen Outhoot—for which the following Patent was given:

Wee Phillip Carterett Esqr Governor of the province of & Supra have given & granted in the names of the Right Hon John Lord Berkley, Barron of Stratton and Sr George Carterett Knt and Baronet the absolute Lord Proprietors of the said province and by these presents doe give & grant unto Fopp Jansen Outhout a parcel of land lying & being between two Creekes on the east side of De Lawarr River over against New Castell called by the Indians Hoppemense, joining on the north side to Derrick Albertsen and on the south side to Mickell La Croix, which said parcell of Land and meadow thereunto belonging does contain according to the survey which is to be made by the Survayor Generall or his Deputy and to be hereunto annexed—yielding & paying to the said Lords Proprietors their heirs or assigns on every 25 March according to the English acct one half penny of lawfull English money for every one of the said acres therein contayned, the first payment whereof to begin on the 25th day of March which shall be in the year of our Lord 1670.

Given under the Seal of the Province the 3rd day of May 1669 in the 21 yeat of his Maj Reigne Charles the Second.

Vol I of Deeds & Patents, page 35

Fopp Out Hout Jansen Purchase

Fopp Out Hout Jansen, probably the first permanent European settler in Salem County, bought his land of Machierick, Hictock, Taspeemick, Wenamink, Keckquenner, the Indian chiefs in 1665.

Purchase by Isaac Tinna a First Settler, Salem County

This deed bindeth me, Wenamick, Sackamacar of certain tracts of Land lying in the south side of Delaware River over against the town of New Castle upon Delaware River, I the afore said Wenamick, Sackamacar of the afore said tracts of land doe acknowledge by these Presents, to sell unto Isaac Tinna of New Castle, one tract of the aforesaid Land belonging unto mee, which is bounded upon the River side from the Creke this side Sobe Johnsen unto the other side of the West Creke of the plantation, for which tract of Land I doe acknowledge that the aforesaid Isaak Tinna hath fully satisfied and paid me for and in consideration whereof I doe give the above said tract of Land unto the aforesaid Isaac Tinna to have and to holde to him and to his heires for Ever.

In Witness whereof I have hereunto sett my hand this 20th of Jully 1666.

Witness hereof
John Carr
Anthony Braiant.
Vol 1 of Deeds, page 21.

The X marke of
Wenamick, Sackamacar

The Lecroy Purchase, Salem County

Wee Tospamink, Wienaminck, Machkierck Allom, Maghaeckse and Keckqueneminck The Natural owners doe hereby acknowledge for us and our children or heirs to have sold a piece or parcell of land containing in Breath and wide as ye same is lying between and behind Foppe Outhout and Machiell Lacroy for ye sum of one gunn, Ten fathom of wampum, three hogs and one kettle which said goods we do hereby acknowledge to have received and thenke fully paid us this 6th of November 1667 in ye house of Fopp Jansen on ye eastern Shoare called New Yernsey. this sale to Peter Lacroy and Machiel Lacroy Jr.

The mark of X Wienaminck
The mark of X Machkierck Allom
From a certified copy owned by the compiler.

Lucas Peterson Purchase

A Translated Dutch Deed from the Indians for Land in Salem County

"Wee, Kerpenneming and Mattein—meke Brothers together declare and know to have sold unto Luycas Peterson, all that tract or parcel of land that he without hindrance may live upon, lying upon ye south side of Swart Hooke, and upon ye north side of the land of Matys Mattsey, and Matys Niellissen right by a line out to the Creek, and that for a good consideration in goods as we have already received to our content and make it from us free of all pretensions or actions, that we can ever have upon the land, the buyer takes now into his possession: this we witness with our marks upon Luycas hooke ye 30th December 1671.
This is witnessed, ensealed and
Delivered to Luycas before me one
of ye Governors Justice of ye
peace in New Castle
Foop Janse OutHoot"

X Mark of Kerpenneming
X Mark of Matteinmeke

72

"A true Coppy out of
Duch written in
English by me
        Tho. Spry"

A license was granted to purchase land over against Christeen kill to Lucas Pieters Sept. 10, 1668. This "true Coppy" is in possession of the compiler.

The Cantwell and Dehaes Purchases

On February 8, 1673, Tospaminck and Weinamink sold to Edmund Cantwell and Johanes Dehaes about seven hundred acres of upland meadow and marsh on the south side of Jeremiah's Kill, as far as Finns Creek along the river at Pompion-hook.

The consideration was one-half ankor of drink, two match coats, two axes, two barrs of lead, four hand fulls of powder, two knives and some paint.

A Certificate Signed by Indians of Salem County

The following certificate is in the possession of the Historical Society of Pennsylvania. No mention of it is made in the Salem deeds of N. J. Archives. Where recorded is uncertain.

It is not a deed, although it is signed with the marks of four Indians of Fenwick's Colony as were the receipts.

"Aprill ye 10 1676 Meopeny Sent his man ton hunt with us hipolit Lefevour Senior and John Pledger to marke ye bounds of our land yt we bought of the ingins we went from our houses and ye point east by South about a mile and a half from our houses there we did marke a white oke from thence South west about 2 miles we did marke a white oke from thence to the head of a Creeke being the second run towards the towne this Creeke doth bound ye west side of our land there we did marke a white oke from thence to ye creeke on ye north west side of ye land at ye head of ye Creek we did marke a white oke from thence north wards to the side of Maneting creeke we did marke a white oke from thence to ye head of a branch of Maneting Creeke we did marke a white oke. This done to ye order of Meopony, Alloways, Accomes, Sacetorus.

                        Meopeny
                        Accomes
                        ton hunt
                        Sacetorus
                            by their marks

In witness of us
Thomas (T W) Watsun   (Wax)
William Willis   (Seal)
        This certificate is endered upon Record
                        Fenwick Adams

Payd to the injins for land for
four famelys
four match coats
one half ancur rum
two staves of lead
two dubble handfulls of powder

Accomes
Meopeny

by their marks

More paid four match coats
two payre of stockings
two knives

Sacetores
Aloways
by their marks

for the land belonging to hipolit Lefever and for the land belonging to
John Pledger
One half ancur of rum
two gunns
two kettles
two looking glasses
two alls
two hoes
two needles
two spoons full of paynt
    Accomes same as Nikomis.

    The spellings are copied from original documents, which accounts for
the many variations.

First English Settlers in Salem County

    Hippolit Lafever and John Pledger arrived in Salem County and
bought land of the Indians before Fenwick arrived on the Griffin on the
23-9 mo. [Nov.] 1675. The first Fenwick Indian deed was dated November 17th.

    John Pledger was in Southern New Jersey as early as March 13,
1674/5. His wife and son came with Fenwick.

The Hendrickson Purchases

    On June 10, 1675, Osawith sole Indian proprietor sold to John and
Peter Hendrickson two necks of land called Singletree or Enboome by the
Christians and Emaijens by the Indians. The consideration was "two
match coats, two double handfulls of powder, two half ankers of strong
liquors, two half ankers of strong beer, two awls, two barrs of lead and
two needles."

74

An Indian Deed of Gloucester County, Recorded in New Castle, Delaware

Asawit, Oppeck Jan and Runnuckle (or Awieham) on Nov. 15, 1676, sold a neck of land opposite Marcus Hook beginning on the West or lower side of a Creeke called Mattieh Cussing by the Indians and Old-mans Kill by the Christians, up the river side to the first small kill called Cachkikana hacking and south east into the woods including all the land between the said creeks.

The consideration paid by Hans Hoffman and Peter Jansen was two half ankers of liquors, two guns, two match coats, four double handfuls of powder, two kettles, four bars of lead, four looking glasses, four knives and four awls. Guns Justasen on their behalf also paid the Indians one gun, one anker of beer, one bar of lead and one double handful of powder, according to the "Records of the Court of New Castle" published by the Colonial Society of Penna. See also 1929 Year Book, New Jersey Society of Pennsylvania, article about New Stockholm.

The Jennings Purchase

"20th 6 mo 1681 Schockanam and Etthunt of Cohanzick sold to Henry Jenings a tract of land containing about 300 acres next to Jacob Youngs fronting up a little creek consideration 2 match coates, 2 mutches of powder, 2 barrs of lead, 2 knives, 6 fish hooks, 6 needles, 20 spoons of red lead and to make a coat or two. Witness Thomas Smith, William Fromes." Recorded Nov. 28, 1682.

The John Nichols Purchase

Shawkanum and Et-hoe brethren and Indian proprietors sold for the consideration of one double handful of powder, two bars of lead, two knives, three penny worth of paint, one hoe, one axe, one looking glass, one pair of scissors, one shirt and one breech cloth on 25th day 4 mo. 1683 to John Nichols of Nicholas Hartford near Cohansey a piece of ground containing one hundred acres. This land was in Catanungut and bordered on the lands of George Hazlewood, Henry Jennings and Samuel Bacon. The deed is copied in full on page 500 of Cushing and Sheppards history.

Abstract of Samuel Bacon's Indian Purchase

Know all men by these Psents yt Wee Shawkanum and Et hoe Indyan proprietors of that pcell of Land called & knowne to the Indyan Natives by the name of Cata-nun-gut lying near Chohanzey on Dellaware River for and in consideration of two coates of Duffelds, three Blanketts, two double handfulls of powder, Six barres of lead, two shirts, two knives, two pair of Stockings, two looking glasses, two combs, two hoes, two axes, two needles, two aules, one Gunn, one gildr in Wampon and two pair of Cicers to us in hand paid by Samuel Bacon Senor of Woodbridge in East Jersey yeom. at or before the sealing and delivery hereof the receipt whereof is hereby acknowledged, have granted bargained, sold aliened enfeoffed and confirmed unto the said Samuel Bacon his heirs and assigns forever, a pcell of Land containing by estimacon four hundred acres (be it more or less) butted & bounded as followeth, Beginning at a fast landing on Chohanzey Creek formerly called Jacob Youngs Neck, now and hereafter to be called Bacons Adventure without the disturbance of us Shawkanum and Et-hoe Brotheren or either of us our heirs or assigns or any other pson lawfully claiming to have any right or Indyan interest therein and thereunto Wee bind ourselves and our heirs joyntly & severally in the penalty of One Hundred Buckskins to be payd by us, our heirs or assigns.

In Witness whereof we have to these psents each Ptye putt our hands & Seales the 25th day of the 4th Moth in the year according to the Englais accompt 1683.

Testis	Signed & Sealed	
Richd Guy	Shawkanum	X his mark
James Nevill	Et Hoe	X his mark
Justices		

Recorded ye 14 of July p me. Samuel Hedge Rec.
Salem 2 of Deeds, page 87

I saw the original of this deed fifteen or more years ago but the owner of it would not permit it to be photographed. I have forgotten her name. Samuel Bacon settled in New England before he came to New Jersey.

Indian Deed for Land at Pensauken

In the history of the Surveyors' Association of West Jersey there is a copy of an Indian deed dated April 12, 1684, to John Roberts, Timothy Hancock and William Matlack for land at Pensauken, an Indian town. The consideration was one match coat, one little runlet of rum and two bottles of rum. This deed was signed by Tallaca and witnessed by Nackontakene, Queieckolen, Notthomon, and others presumably white men.

Cinnaminson and Oneanickon were also Indian towns.

George Hutchinson Purchased an Island Near Morris River

This Indenture made ye Tenth day of ye fifth month July in ye year of or Lord one thousand six hundred ninety & four Between Mr. John Indian King of Stepsons Island in West New Jersey of ye one pt and George Hutcheson of Burlington in ye Province of West Jersey aforesd merchant of yet other pt.

Witnesseth that for & in consideration of ye sume of Twenty pounds Currt money of this Province of West Jersey, to ye said Mr. Thomas in hand paid by ye said George Hutcheson ye receipt whereof he ye sd Mr. Thomas doth hereby acknowledge & thereof & every part thereof & pcell thereof doth acquit & exonerate release & discharge ye sd George

76

Hutcheson his heirs Exrs & Admrs & every of them forever by these psents Have granted bargained & sold alyened & enfeoffed & confirmed and by these psents doth fully clearly & absolutely grant bargain & sell alyen enfeoffe & confirm unto ye sd George Hutcheson his heirs & assigns forever a certain Tract of Land called Stepsons Island lying betwixt Cedar Hamake & Morrises River fronting ye Bay of Delaware River & bounded with two brooks on each side called by ye names of Semes & Toyoken & from ye mouths of ye sd brooks to ye heads of ye same.

In Witness whereof ye sd pte first above named by this psent Indenture hath sett his hand & Seale ye day & year first above written 1694.

Master Thomas by his mark & seal.

Signed & sealed in ye
presence of
Richard Jones
Eustame Jones by her mark
Danll England
Vol B, part 2, pg 617 of Deeds. Sec. of State of N. J.

Mr. John and Master Thomas seem to have been confused in this deed.

Stepsons or Stephants was probably Stephens as the same is now applied to a creek tributary to the Great Egg Harbor River.

Indian Gift of Land

"Mauhauxsey and Cuttanoque July 2, 1694 all the tract of land lying and being betwixt ye head of Cohansey and Maurises River partially for the love we bear unto Peter Erickson of Cohansey for all the service he has done us as interpreter betwixt us and our brethren the English in all Treaties that we have had occation for him.

"Do give and grant unto said Peter Erickson all ye tracts or parcel of land lying and being ye plantations of Leonand Bereman & Richard Whittaker bounding between Cohansey & ye Back Creek being excepted out of that tract of land sold by us to Jeremiah Bass on acct. of the West Jersey Society.

"Witnesses Jeremiah Basse,
Jos. Barkstead, Tho. Bridge."

Mohocksey Wanted Peace

At a meeting of the Provincial Council held at Philadelphia, July 6, 1694, it was reported by several Delaware Indians, among whom was Mohocksey, the celebrated Chief of Salem County "that the Onondages and Senekaes had sent to them claiming that they staid at home and boiled their pots and were like women," while the other two nations went abroad and fought against the enemy. They wanted the Delawares to fight against the French. Mohocksey said, "The former belt sent by the Onondages and Senekaes is sent to us all and we have acquainted one another with it, and tho we live on the other side of the river yet we reckon ourselves all one because we drink one water. We have had a continued friendship with all the Christians and old inhabitants of this river since I was a young man and are desirous to continue the same soe long. as we live." Tamanee and other Indians were also anxious for peace.

Mohocksey was the ranking chief or king of the New Jersey Indians of the lower Delaware River section. He is certainly worthy of some memorial at the hands of the descendants of those he termed Christians. His lofty ideals far exceeded those of the Europeans. On the highest hilltop, in Old Salem County, a statue of grand old Mohocksey should be placed, facing the West, where he saw the sun sink below the horizon.

The Samuel Hedge Purchase

On December 2, 1699, Cottenockque, Aweham, Casecoping, Kishwigwom, Quishittue made a deed to Samuel Hedge for a piece of land beginning at Hance Oulsons bound tree on Oldmans Creek, thence up said creek to line of Thomas Piles ten thousand acres called Pilesgrove and so north east to Oldmans Creek and down ye several courses of ye creek to said Oulsons.

The consideration was twelve match coats, two gallons of rum, one anker of cider and divers other things not mentioned. The witnesses were Benjamin Acton, Elston Wallis, William Champneys, John Nicholds, Jeremiah Smith, Simon Morgaine and Samuel Hedge, Jr.

William Evans Purchase

On the 6th day of ye 8th month 1701, William Evans, of Burlington County, purchased of Hicmorn, an Indian chief, one thousand acres of land for a consideration of five pounds. This deed is apparently an unusual one, in several ways, but the writer was unable to go to see it or obtain a full copy of it. The tracings obtained, of the names of the Indians and their marks, are somewhat illegible and without the writing of the deed for comparison it may be that the Indian names are somewhat incorrect. The deed was witnessed by Jonathan and Thomas Eldridge, Opa Hoge, Hokgo Sooway and Getrus Wirker.

The deed is in possession of Joseph S. Evans, a probable descendant of William, the settler.

The Test Purchase

"June 16, 1703, Okeman, Quiateman and Awishman, alias Capt. John sold land to John Test on Hugh Hutchings run on line of Thomas Graves and Aquilla Barber and Salem Creek. Consideration two pounds, one piece of eight, one quart of rum and several other valuable considerations.

"Witnesses, Sam'l Hedge, Sam'l Hedge, Jr., Jon Hopman, Jon Child, Jon Dickason, Jon Smith, Charles Angelo, Benj. Dewall, Jon. Holme."

A Purchase by Cape May Settlers

Know All Men By These Presents Righting That I Panktoe have sold unto John Dennis and his associates, that is to say Samuel Mathew, Samuel Crow, Joseph Whilden, Ez Eldredge, John Carmon, James Pittny, a sarting tract of land and meadow for a sarting sum of money in hand already received the bounds as followeth, that is to say from the creek and so running up to the woodland along by Carman's line to a white oak tree at the head of the swamp, the trees being marked on all four sides by a pond joining Jonathan Pines bounds all our land and marsh lying and being between the bounds above mentioned and the Cape Island. I Panktoe have sold with all privilege their unto belonging from me and my ares forever to their heirs forever whose names are above written, to have and to hold and peasable to enjoy without any molestation from me or any other by me pretmement.

Given under my hand this 5 January 1687.

Witness

Abiah Edwards   Panktoe his X mark

John Carmon

Recorded Nov. 2, 1814 Libra AV of Deeds, page 526 in the office of the Secretary of State of N. J.

James Lime, Secy.

This was taken from a copy in possession of the Cape May County Historical and Genealogical Society. Mr. R. L. Goff, former Cape May County Engineer has a copy of an unrecorded Indian deed dated August 15, 1702 for land in Monmouth County.

Deed of Indian "King Charles" of Little Egg Harbour Ratified

"Bee it known unto all people whom it may, shall or doth concern; That Where as In the year according to the English account in the month of Agust, one Thousand six hundred seventy & four; one Nanacuttun owner of a certain tract or parsell of land down at the sea side on the westerly side of the bay or sound and Joyning upon ye north westerly side of little Egg harbour river; comonly called & known by ye Indian name Amintconk; Having for divers causes & considerations him There-unto moveing but more especialy the consideration of ten pounds In Cur-rant money of New York or the value thereof in ten match coats and other Indian goods paid unto the Above said Nannacuttun by one Henery Jacobs Faukinburge Inhabitant upon mattiniconk Island in delaware river; given granted infoefed & confirmed unto the above said Henery Jacobs Faukinburge all the above said certain tract or parsell of land & medow adJoyning to the bay or sound & little Egg harbour river & run-eth up into the upland as farr as there is any good land or rather bound-ed upon the barrons; & so farr in breadth as in runing dow again to the bay or sound it doth take in five Islands in the medow or marsh 3 great ones & two lesser ones: unto him & unto his heirs for ever more; with all the reversion & reversions, remainder & remainders with all

79

Isues & profits of all & singular ye premises and every of their apur-
tainances before mentioned to be granted unto the said Henery Jacobs
Faukinburge & his heirs to the use of the said Henery Jacobs Faukinburge his heirs & assignes for ever more To have And To hold the said
land medow & Islands in the same as above said unto him the said Henery Jacobs Faukinburge & to his heirs & asignes for ever more; Now
know yee I mahomecun Alias King Charles Couzen unto the above said
Nanacuttun he being my unckle do hereby grant Ratify & confirm all
whatsoever Nanacutton my unckle hath don in & for the premises; And
further for the said Henery Jacobs Faukinburge satisfaction I do hereby
convenant & promise that he shall fully freely & peacably have hold &
Injoy the above said tract or parsel of land medow & Islands in the same
as above said with all & singular the premises above mentioned to be
granted by nanacutton my unckle And (not only to grant) the above
said land and premises bounded as above said as may or doth apear by
marked trees marked by Indians acording to order; but also do further
promise and ingage to maintain warrant & defend the said land and
premises unto the said henery and to his heirs & asignes for ever more;
& to the utmost of my power will maintain warant & defend the same
against all persons whatsoever claiming or to claim from by or under
nanacutton my unckle or from by or under mee And I do further declare
that the above said tene pounds paid as above said for the land above
said to be the full satisfaction and valuation for ye above sd land &
premises; the receipt where of is hereby owned to be fully content &
paid; And In wittness to the truth of all & singular ye above mentioned
premises I have hereunto set my hand & seall This being the Eleventh
day of Aprill one Thousand six hundred ninty & seavn in the presants
of other Indians &t:"
Testis
Signed and sealled and delivered in the presants of us
Daniel Leeds, Justice
William Leeds, Seanior vietor

leneaquah
his mark
quetayaha
his mark
mahamecun
his mark
Soswoaton
his mark

May ye 3rd 1699

Recorded ye above written deed in the Publick
Records of the Province of West new Jersey:
sset. 644 & 645: libe 33 p.m. Thos. Revell Survey-
or and Regt.

The above deed is in possession of Mrs. John M. M. Dudley, Elizabeth, N. J. Copy furnished by Charles A. Philhower.

Extract of a Stray Indian Deed of Monmouth County

Know all men by these presents that We Barnigat Pam, Indians Neell, Whscatme, Squan kink & stor with several other Indian Sachems above mentioned lying in the County of Monmouth and province of East Jersey in America for and in consideration of a certain considerable sum of moneys value to us in hand payed by Gawine Drummond of Loch arbor in the said countie and province the receipt of which we doe hereby acknowledge and ourselves therewith fully satisfied contented and payed and of and from every part and parcell thereof do by these psents clearly and absolutely acquit exoner and forever discharge the said Gawine Drummond his arres, exers, administrators and assigns or any or either of them forever.

Have aliened, granted bargained sold in feoft and confirmed, unto the said Gawine Drummond and his friends and any of them forever All the two tracts of land and meadows on the branches of Manasquan called and known by the names of Mochipinos and Matacharson which tracts of land and meadows is bounded all round by our other lands. Which two tracts of lands and meadows called Mochipines and Matackarson and premises above mentioned we the said Pam, Neell, Whiscatine, Storo and the rest of our partners Binds and oblidges us our aires and successors to warrant secure defend and harmless keepe said Gawine Drummond.

In witness whereof We the said Barnigat Pam Indian, or Neell, Whiscatme, Squankunk, Store with the advise and consent of svall more Indians concerned have hereunto sett our hands and sealls at Locharbor the ffourteen day of August One thousand seven hundred and two 1702

Signed sealled and delivered in pnt of	X Pam	Seals
mark Robert R. D. Drummonds	X Neell	Seals
mark George G. R. Roger On Reverse side	X Whiscatine	Seals

April ye 6 day Annoqdomimi 1722 George Rodgers one of the within subscribing evidences took a sollom oath upon ye Holy Evangelislos of Almity God that he saw two of ye within named Indian Sachems sign seal and deliver this within written instrument as there act and deed for ye use within mentioned and received ye certain considration money

Henry Leonard

(Above deed was copied from a facsimile tracing July 10, 1931, made by Surveyor R. L. Goff, Dec. 27, 1898, of original deed then among the Pines in Atlantic County. It is herewith published in the belief that it was not recorded and should be made available for interested persons.

Committee of Indians Given Power of Attorney to Dispose of
Remainder of Their Lands

"Know all men by these Presents that we the subscribers Teedyc- scunk, King of the Delawares, George Hopayock, from the Susquehannah, Ben Claus, Jo Wooley, Josiah Store, James Calvin, Peter Calvin, Dirick Quaquay, Ebenezer Wooley, The widow of Quiquahalah, Sarah Store to whom their respective Husbands had given their estate, Andrew Wooley, George Wheelright, Joseph Cuish, Will Loulax, Gabriel Mitop, Zeb Conck-kee, Bill News, John Pombolus, Tom Evans, Robt Keekott, Jacob Mullis, Abraham Loques, Isaac Swauela, Indian Inhabitants of the Province of New Jersey having on the twentyeth, twenty first, twenty second and twenty third days of February one thousand seven hundred and fifty eight at Croswicks met the Honourable Andrew Johnston and Richard Salter Esqs Charles Read, John Stevens and William Forster Esqs Commissioners of the Colony of New Jersey appointed by Law to enquire into the claims we or any of us have to Lands in New Jersey, Have come to a resolution to Impower a Committee to transact the Business which may be necessary respecting the Land We or any Indians in New Jersey may have and claim within the same, and having this day delivered to the Commissioners in open Council a full and ample List of Lands which we or any of the Indians of New Jersey claim, the minitts of which Council containing such List, signed by the Commissioners or the major part of them and the attornies or Committee herein appointed or major part of them, We hereby declare to be the full of our claims and Demands, and full and conclusive evidence thereof, We do for ourselves and our Heirs and for all the Indians claiming or pretending claim in New Jersey Release to the General Proprietors of the respective Divisions and to the Purchasers under them all other Lands with in the same not particularly specified in the said List or account and all other Lands in the said List for which Indian deeds executed by the Indian Inhabitants shall appear, excepting the Rights of the Minisink and Pomton Indians to any Lands in the Northernly parts of the Colony, and we do by these presents constitute and appoint Tom Store, Moses Tottamy, Stephen Calvin, Isaac Stelle and John Tompshire our Committee or attornies to transact all manner of business with the Commissioners appointed or to be appointed by the Governor or Legislature of New Jersey relative to the Lands we or any Indians in New Jersey may claim and have not sold in the same, hereby fully authorizing and impowering the said Tom Store, Moses Tottamy, Stephen Calvin, Isaac Stelle and John Pompshire or the major part of them to come into any agreement concessions and resolutions respecting the Indian Lands in New Jersey, and in their own names to give and execute full conveyances acquittances and discharges for the whole or any parts thereof ratifieing and confirming whatsoever our said Committee or attornies or the major part of them shall do in the Premisses.

In Witness whereof we have hereto set our hands and Seals at Croswicks this twenty third day of February in the year of our Lord one thousand seven hundred and fifty eight.

Teedescunk X		Peter Calvin X	Seal
King of Delawares	Seal	Andrew Walley	Seal
George Hopayock X	Seal	George X Wheelright	Seal
Ben X Claus	Seal	Joseph X Cuish	Seal
Josep Wooly	Seal	Will X Loulax	Seal
Josiah Store	Seal	Peepy X his mark	Seal
James Calvine	Seal	Philip X his mark	Seal
Diric Quaquay	Seal	Saml X Gosling	Seal
Gabriel X Mitop	Seal	John X Pombolus	Seal
Zeb X Conchee	Seal	Tour X Evans	Seal
Will X News	Seal	Robert X Kekott	Seal
Ebenezer X Wooley	Seal	Jacob Mullis	Seal
Sarah X Store		Abrahm X Loques	Seal
widow of Quiquehala	Seal	Isaac X Swauela	Seal

The above twenty eight Indians sealed and delivered the above Power of Attorney on the Day, month & year therein Specifyed in presence of us the Subscribers the three last arriving after the Power was delivered up.

Wm Tennent
Jacob Lawrence
Abel Middleton
Amos Middleton

We Tom Store, Moses Totamy, Stephen Calvin, Isaac Stelle and John Pompshire the above Committee appointed by the Above Constituents do for ourselves and our Heirs agree to every article clause and thing contained in the foregoing writing or Power of Attorney. Witness our hands & Seals at Croswicks the 23 Feb 1758.

Wtinessed by

Wm Tennent
Jacob Lawrence
Abel Middleton
Amos Middleton

Tom X Store	Seal
Moses Tomami	Seal
Stephen Calvine	Seal
Isaac Still	Seal
John Pumpshire	Seal

1-2 of Deeds, page 45.

The End of All Indian Rights in the State of New Jersey

"This Indenture made the seventh of April in the year of our Lord one thousand eight hundred and thirty two (1832) Between the party or Tribe of the Delaware natives of Indians, formerly residents of the southern part of the State of New Jersey, and now located and residing at or near Green Bay in the Territory of Michigan, party of the first part, and the State of New Jersey of the second part.

Whereas in and by an Act of the Legislature of the said State of New Jersey, passed on or about the twelfth day of March last, entitled,

"An Act for the extinguishment of every right, title or claim, which the Delaware Tribe of Indians, formerly residents of New Jersey, and now located at Green Bay, in the Territory of Michigan, now have or ever had, to any part of the territory of New Jersey or its franchises," it is recited and enacted as follows, to wit, "Whereas the Delaware Tribe of Indians, formerly residents of New Jersey, and now located at Green Bay, in the territory of Michigan, have memorialized the Legislature of this State, 'setting forth that in the respective treaties, deeds and conveyances, whereby the lands south of the river Raritan were ceded and transferred to the State of New Jersey, the right of said Tribe to the fisheries in the rivers and bays of said State, south of the river Raritan, was reserved, and has never relinquished or alienated, which fisheries are now used and possessed by the citizens of this State; and have authorized Bartholomew S. Calvin, a Chief and principal member of said Tribe, resident at Green Bay, aforesaid, to lease sell or transfer said fisheries, and to receive such compensation for the same, as this Legislature may deem proper to grant; and Whereas it is represented, that the legal claims or title of said Indians to the fisheries aforesaid, are barred by reason of their voluntary abandonment of the use and occupancy of the same, but that this legislature should grant a remuneration for the right of said fisheries, as an act of voluntary justice, as a memorial of kindness and compassion to the remnant of a once powerful and friendly people, occupants and natives of this State, and as a consummation of a proud fact in the history of New Jersey, that every Indian claim, right and title, to her soil and its franchises have been acquired by fair and voluntary transfers—

Therefore—Sec. 1, Be it enacted, by the Council and General Assembly of this State, and it is hereby enacted, by the authority of the same, That the treasurer of this State, for the time being, shall pay to the aforesaid Bartholomew S. Calvin the sum of two thousand dollars, as soon as the said Bartholomew S. Calvin shall make and file in the office of the Secretary of this State, such deed or other instrument of transfer, which shall be approved by the Governor of this State, as a good and valid conveyance and transfer in the law, to the State of New Jersey, of all the soil, fisheries, or other rights, or reservations which now or ever were owned or possessed by the aforesaid Delaware Tribe of Indians, to any portion of the territory of New Jersey" as by reference thereto will appear—

Now this Indenture witnesseth, That the said Party or Tribe of the Delaware Nation of Indians, for and in consideration of the sum of two thousand dollars, legal money of the United States by the Treasurer of the said State, (for and in behalf of the said State of New Jersey) to the said Bartholomew S. Calvin in hand paid (to and for the use and behoof of the said Party or Tribe of the Delaware Nation of Indians) the receipt whereof as well as the said Party or Tribe of the Delaware Nation of Indians as the said Bartholomew S. Calvin, do hereby acknowledge, have granted, bargained, sold and transferred, and by these presents, do grant, bargain, sell, alien, enfeoff, release, convey, assign, cede, transfer and confirm unto the State of New Jersey, successors and assigns, all the soil, lands, bars, bays, rivers, and waters, whatsoever wtihin the bounds, limits or jurisdiction of the said State of New Jersey; Together with all and singular the fisheries, fishing, fowling, hunting, commons, franchises, royalties, rights, members, privileges, hereditaments, and appurtenances thereunto belonging or in any wise appertaining, and the reversions and remainders, rents, issues, and profits thereof —Also all the estate, right, title, interest, use, property, possession, claim, and demand, whatsoever, either in law or equity of them the said Party or Tribe of Delaware Nation of Indians, or any of them, which now are, or ever were, owned, possessed, used or occupied, by them, or any part or party thereof, of, in and to the said State or Territory of New Jersey, or any part or portion thereof or of, in and to the fishing, hunting and appurtenances as aforesaid, of, in and appertaining to the same and every part thereof, To have and to hold unto the said The State of New Jersey, successor and assigns, To the only proper use; benfit and behoof of the said The State of New Jersey, successors and assigns forever.

In Witness whereof Jeremiah Johnston, Charles Stephens, Austin Quinney, Sampson Marquis, Andrew Miller, and Bartholomew S. Calvin, Chiefs and Principal men and representatives of the said Party or tribe of the Delaware Nation of Indians, for and in behalf of the said party or tribe, have hereunto set their hands and Seals the day and year first above written.

Signed, Sealed and delivered in the presence of us; The said Jeremiah Johnston, Charles Stevens, Austin Quinney, Sampson Marquis and Andrew Miller, by the said Bartholomew S. Calvin, in their names, as their Attorney, and as their Act and deed, by virtue of a Power or authority (hereunto annexed) enabling him thereunto, hath hereunto set their hands, and seals the day and year aforesaid.	Jeremiah Johnston	Seal
	Charles Stephens	Seal
	Austin Quinney	Seal
	Sampson Marquis	Seal
	Andrew Miller	Seal

By their Attorney
Bartholomew S. Calvin
Bartholomew S. Calvin

On seventh day of April 1832 before me Thomas Gordon one of the Masters of the Court of Chancery, in & for the State of New Jersey, appeared Bartholomew S. Calvin and acknowledged that he signed, sealed and delivered the above, in the name of the persons as their attorney. Approved and signed at Trenton 10 of April 1832.

P. D. Vroom
Governor of New Jersey.

Trenton, N. J., April 10, 1832.

Received of the State of New Jersey by the hands of Charles Parker, the sum of two thousand dollars, being in full of the consideration mentioned in this deed.

Witnesses present   Bartholomew S. Calvin
Wm. Hyer
John McKelway

Mis. No. 11

All Indian Lands Purchased of Them

It is a historical fact that every acre of ground in Southern New Jersey was purchased of the Indians by the Dutch, Swedes and English. On page seventeen of Volume Twelve of the "New York Historical Records," devoted to the Colonial Settlements on the Delaware River, there is a copy of a patent for land in Cape May County. The Indians involved were Sawowouwe, Wuoyt, Pemhake, Mekowetick, Techepewoya, Mathamek, Sacoock, Anehoopen, Janqueno and Popahake. This patent dated June 3, 1631, to Samuel Godyn and Samuel Bloemmaert, called for a tract of land, four (Dutch) miles long and four miles wide, or sixteen square miles. The consideration is not mentioned. On page forty-nine is a copy of an Indian deed for a tract of land between Rankokus Creek and an unnamed creek on the south end of Tinnekonk Island (Burlington, N. J.) including said island and other lands. This deed to Symon Root, Alexander Boyer, Peter Harmsen, David Davidsen and Cornelius Mouritsen was dated April 9, 1649. The Indians who signed the deed were Kickeesickenom, Hattowens, Kintakosy and Schinna.

During 1641 the Swedes purchased the land between Raccoon Creek and Cape May, of two Indians named Wickusi and Mekopemus. The English had previously bought some land on Salem Creek but it was claimed the Indian who sold it did not own it. In 1649 the Swedes bought another tract between Raccoon and Mantua Creeks. For further information about the Swedes and their dealings with the Indians the reader should consult "The Swedish Settlements on the Delaware," "Lindestrom's Geographia Americae," and "Instructions for Johan Printz," all written by Amandus Johnson and published by the Swedish Colonial Society.

Governor Andross and the Magistrates May 13, 1675, at New Castle, Del., renewed a treaty of peace with Renowewan, of Sawkin; Ipan, of Rancocuskill; Ket-marius, of Soupnapka, and Manickty, of Rancocuskill. The Indians were given some belts of wampum, four match coats and four lappcloathes, for which they returned their thanks and canticoed.

Abstract of a Deed to Daniel Coxe in Northern Part
of New Jersey

"To all people to whom this present writing shall come, Wee Hoe-han, Kepanockonickon, Romasickamen, Tipaopaman, and Vevenutting, Indian Sackamackers and owners of ye Tract of Land herein after particularly named within the Province of West Jersey send Greeting—

Know yee that the said Indian Sackimackers and owners of ye said Tract or Tracts of Land herein after menconed for & in consideration of one hundred fathoms of Wampan, Tenn stript Matchcotes, Ten coates, Tenn Gunns, fifteen kettles, forty matchcotes, forty strowd water matchcotes, twenty shirts, forty pair of Hose, thirty howes, two anchors of powder, one hundred knives, one hundred barrs of lead, sixty pounds of Shott, two barrels of beere, three pounds of Red Lead, three hundred pipes, three anchors of tobacco & two half anchors of Rumm & five matchcotes, By Adlord Bowde now of Burlington in the said Province of West Jersey merchant for and on ye behalf of Daniell Coxe Esqr Governor of and Proprietor in the said Province at and before the sealing and delivery hereof to them the said Indian Sackimackers and owners well & truly in hand paid whereof and wherewith they dow hereby acknowledge themselves fully contented & satisfyed, Have granted bargained & sold alyened enfeoffed & confirmed, and by these psents doe fully clearly and absolutely grant bargain & sell alyen enfeoffe and confirm unto the sd Adlord Bowde for and to the only proper use and behoof of the said Daniell Coxe and of his heirs and assigns forever, all that and those Tract and Tracts of Land laid forth and markt and bounded as hereafter followeth, That is to say, beginning at a point upon Thomas Budds Lyne looking northwesterly and from thence by a Lyne of marked trees running north somewhat westerly to ye northmost branch of Rariton River, and thence down the sd River so low as to a road leading from Dellaware ffalls towards Yorke by Vincents plantation and from thence southerly along the said road untill it meets with another of Thomas Budds Lynes passing over the Stony Hills upon ye east side of Milstone River, then from thence upon ye said Lyne northwesterly to a corner tree, Thence west southerly on ye sd Lyne to a corner tree, & lastly from thence south westerly to the aforesaid first poynt."

"In Witness whereof the said Indian Sachimackers and owners have hereunto sett their hands and affixed their seales (according to English acct) the nyneth day of the moneth called Aprill in the year of our Lord one thousand six hundred Eighty and Eight and in ye fourth year of the Riegne of our Lord King James the Second over England &c."

Signed Sealed & delivered		Hoeham
in presence of	Henry Jacobs Fulcon-	Kepanoockonickon
Andrew Robison	bridge interpretter.	Romasickamen
Thomas Budd	Awshoppa an Indian by	Tiptaopaman
Tho. Bowman	his mark.	Vevenutting
Henry Grubb	Vol B part 1 of Deeds, page 181	by their marks.

Noted Lenni Lenape Indians

(By Bessie B. Warwick)

Niconique ruled the Unalachtigo Tribes in and about Salem.

Himolin ruled between Haddonfield and Upper Evesham.

Arasapha, who sold Pine Point, Camden, to the Cooper settlers and who made the peashore trail famous.

Eagle Eye, who ruled south of Haddonfield.

Carlyle, who ruled the Tribes in the entire Crossweeksung Tracts to beyond the Raritan.

Tamany, who was a Delaware Chief with territorial jurisdiction in the Minisink countries of New Jersey, and who was living as late as 1690, was the greatest Chief. He was said to have never had his equal. Indians are very reluctant to speak of their dead. The ancient writer from whom I gathered the foregoing information said, "Tammany—Our Greatest—threw his influence all about us as he reconnoitered the Delaware Valley." The peashore trail which began below Bordentown was very familiar ground upon which he softly travelled in his deer skin moccasins.

The Indian took his family name from the female line.

Ashatama was an ancient and very honorable name among the ancient Algonquin Indians. This name existed among the Egg Harbor Indians. Ashatama was the name of one of the families who never left the vicinity of the Reservation. The last family of this name was Chief Elisha Moses and his wife, Margaret, who was a strict Quakeress. They lived at the time of the War of 1812, near Egg Harbor. Elisha Moses went on board the "Chesapeake" during that war and was away five years. Margaret gave him up for dead and in the meantime married a mulatto, one of the manumitted slaves' descendants. When Elisha reached home after his very perilous journeys, he took charge of his own household and pre-emptorily dismissed Margaret's mulatto husband; but he assumed the responsibility of Frankie Joel, the son of Margaret's second marriage. Ann Ashatama was his own daughter.

Ann Ashatama was married two times. Her first husband was Peter Green and her second husband was John Roberts.

John and Ann Roberts were the parents of six children. They were named: John, Peter, Richard, Lydia, Hester Ann and Maria.

I remember Maria very well. She died in 1910. She married a man named Marshall, who was lost (after her death), in the swamps for two years. His remains were found just shortly before the World War, 1918.

Both of Ann's husbands were manumitted slaves who had been under the patronage of the Quakers a long time before the Civil War. One of her sons was in the War of 1861. Through the assistance of the people of the neighborhood, she obtained a pension for him. With this money she built a small cabin and lived in it until her death, in 1894. She was buried with some ceremony. The services by the white people of Tabernacle and Shemung were conducted in the old Methodist Episcopal Church at Tabernacle. Her remains were interred in the white burying grounds of that Church. The Clerk's Records will indicate the place of burial. Because of Ann's pure Ashatama blood, she is known as the last Indian of the First Reservation.

The last head chief of the Edgepelick Reservation was Job Moore.

On the Pennsylvania side of the Delaware Valley, Indian Hannah, of near Westchester, who died in the Chester County Public Home, a short time before Ann Roberts, was the last pure blooded Indian of Pennsylvania. She was extolled in song and story by local poets.

Indian Trails in Southern New Jersey

By Bessie B. Warwick

The general trend of the northern and eastern trail of New Jersey either merged with or was a continuation of the great Mohawk Trail in the State of New York.

The Mohawk Trail was joined by a trail along the Raritan River, going southeast, it skirted Matawan, south to Freehold. Thence it followed the Manasquan River to Manasquan. Thence along the Squan Beach and Barnegat, passing near Manahawkin and what is now Tuckerton (then Mathis Island), to Great Egg Harbor Bay, circling Absecon and rounding the point at Cape May.

A trail ran from Burlington to Old Gloucester (Arwamus of the Indians), where it became an upland and a close lowland shore trail. It veered and crossed the creek just above the former American Brown Boveri Shipyards at Camden, near Gloucester. It wound in and out along the creeks and streams to about what is now Euclid Street in the city of Woodbury, where it joined a trail that eventually took it to Salem in one direction. The separation was at the Forks of the road in lower Woodbury. The east branch went along the creeks to Mantua, thence southeast to Swedesboro, south to Woodstown and southwest to Salem. It then followed the river around to the Cohansey Creek; thence along the bay until it reached the shore at Cape May Point.

The west branch of the trail that separated at Woodbury went westward, and crossed the W. J. R. R. (local Delaware River Road), at Lincoln Street. This trail passed Mt. Royal and continued to Clarksboro. It passed directly in front of the Clarksboro Railroad Station, then veered and crossed the main street and followed back of the old Weatherby farms (now the Eglington Cemetery), to below Mickleton. It reached its destination a few miles below Mickleton, where there was a very good sized Indian village half way between Mickleton and Mantua, previous to the Dutch occupation of New Jersey.

There was a short trail at Trenton connecting with one at Princeton. These both joined the Crossweeksung Trail.

Another trail started below Berlin, just above Blue Anchor. This trail crossed the present highway at right angles with this state highway. It followed up the old pike, at present the Camden and Atlantic Railroad, to Long-a-Coming to the forks of that railroad with the late junction of the Camden, Marlton and Medford Railroad into Haddonfield. There it crossed the present King's Highway. It then took a southeast trend and became a land and portage trail to Gloucester, where it joined the Salem Trail.

The Indian Trail to Edgepelick (The New Jersey Indian Reservation) led down from Crossweeksung to Burlington; thence to Mt. Holly (Bridgetown) to Medford (then Upper Evesham) to Indian Mills. This was a land and portage trail.

As late as 1797 the Main street of Medford, which was located much farther to the west and followed the winding streams after it crossed Haines Creek, which crossed the main street of the present day at the North Main Street concrete bridge, was a wagon road following right in the path of the Mantas Atsion Indian Trails from Mt. Holly to Atsion. This trail continued from Atsion to the shore. There it branched one way to Manasquan, and in the opposite direction to Egg Harbor and Leeds Point.

The short trail was from Medford to Tuckerton. It wound through the Flyatt sections of Shemung and Atsion.

The Flyatt Trail branch ran from Small's Tavern to the present Red Lion. Then it went back of Vincentown (then Quakertown) and came out to the road leading to Mt. Holly at Creamer's Corner.

The most notable dry trail was across the state from the Delaware River. This trail touched the site of Camp Dix at Wrightstown. Its general trend was east and west, deviating in the case of streams.

In 1755 Richard Smith, of Burlington, who was the King's Surveyor-General of New Jersey, when on his trip to the Cooper Tract in New York to survey it, used the trails as his mode of travel and thoroughly described the same in his "Story of the Four Rivers."

Branch Street, Medford, was the Cotoxin Branch Trail going east from the Mantas Trail to Quakertown.

The great Crossweeksung Trail led down from above Wrightstown to Mt. Holly.

The Salem Trail, beginning at Arwamus (Gloucester), deviating by reason of the streams to Haddonfield, can be closely followed at this writing by the Old King's Highway to Salem. At Woodbury it came in east of Broad Street, between Broad Street and Green Avenue. It crossed to Broad Street just above Mantua Avenue.

The Cohansey Trail circled Bridgeton and had its side trails known as the Indian Paths, to the Cohansey Trail.

The Assinpink, Rancocas and Cinnaminson Trails all led into old Burlington Pike and adjacent means of entering that city. The Rancocas Creek had double the trails that it had branches running into the main stream. There were trails on the lowland and highland. The main trails being on the warm or southern side of the streams. A few of these trails are still discernible in Medford. One lowland and upland trail beginning back of the Baptist Cemetery crossing branch at the side of the last house before you reach the bridge passes through the Ellis Lippincott Meadow Lands, follows the creek until it reaches the railroad concrete bridge, where it loses itself in the marshlands.

In the Flyatt district a small trail skirts the old dirt wagon road from the North Flyatt Trail and comes in back of Camp Ockanickon.

The old Cotoxin Trail led down from Burlington by the trail from Mt. Holly to Atsion. It turned to the left at Brimstone Corner, which is about a mile north from Main Street, Medford, to the Cotoxin fords. It was known as the Quakerson Trail. It is now a dirt road.

When Captain Mey rounded the cape at Cape May he traveled on several of the ramifications of the Unalachtigo Trails.

Yes, from Trenton to Ocean Beach is the shortest distance across the State. But that east to west trail was a dry trail, hence it avoided all the streams.

From Camden to the shore the trail was a lowland and portage trail. This trail and one in North Jersey were the most traveled trails in the State.

The North Jersey Trail began at Old Bergen (Jersey City). It crossed the Newark Bay, then followed the railroad to Long Branch. There it turned South to Point Pleasant, then it turned almost directly west to Crossweeksung, where it followed to Burlington and then below it joined the peashore trail into Camden. The peashore trail began below Bordentown and ended below Kaighn's Point, Camden.

A lowland trail followed the right side of the Camden, Marlton and Medford Railroad from the trestle bridge east of Haddonfield for some miles. It was around the land that was owned by the great King Himolin.

Our famous South Jersey Indian Trail was from Mathis Island (now Tuckerton), to Mt. Holly. It was the Mantas Trail.

Inian's Ferry, now New Brunswick, centralized all the great inner trails of the State. From these trails eventually nine or more important roads led in and out of the city. Of these many trails the Raritan, Piscataway and the Matchaponix Trails were the most important.

The local Raritan Trail is now the converted towpath of the Raritan Canal.

Indian Beacon Lights

These lights continued down the Palisades at the highest point, always. Then, when the mountains sunk to hills, the hills that were highest were used. All the great trails had them.

The triangular beacon lights indicated the Council Places of Meeting.

The Great Cohansey Council Tribes met at the places now known as the Tumbling Dam at Bridgeton.

Trenton was the Council Meeting place of all the Indians south of the Raritan country and took in the counties of Mercer, Ocean, Burlington and a small portion of Middlesex.

Mt. Holly mount had a beacon light; likewise there was one on the left hand side of the road going from Haines Mills (now Cotoxin Lakes) to Vincentown. It was just beyond the second bridge. Two very large ones were kept burning on the two hills between Medford and Shemung, while several ran up and down the watershed of New Jersey.

The great Council Meetings of all the tribes along the lower Delaware Valley of New Jersey were held at Smith's Island in the Delaware, long since excavated to clean the channel opposite the City of Camden. While Pine Point, Camden, held Council Meetings alternating with the Smith Island Delegations, Tamany presided over these last-named meetings up to nearly 1690.

A very famous Indian Trail from New York to Philadelphia to the Chesapeake Bay was generally used by the Lenni Lenape Indians during the Swedish occupation before William Penn laid out Philadelphia, maps of the same having been made by Col. Henry D. Paxson, President of the Swedish Colonial Society of Pennsylvania, through whose courtesy we have located the trail.

This is the only ancient map of Pennsylvania or New Jersey on which an Indian field was noted. The field, according to Lindestrom, was used so long it was too impoverished to grow corn. It was afterwards a part of Penn's Manor and Penn wrote his manager, James Harrison, to fertilize it well.

The trail ran from New Amsterdam (New York), across New Jersey to the Falls of the Delaware (Trenton), thence on down through Wicaco (Philadelphia), to the Swedish settlement on Tinicum Island and to the head of the Chesapeake Bay region.

The Indian field was on the west side of the Delaware, nearly opposite Trenton, in the old Indian district of Sandhickon. The original map was probably made about 1681. It shows Salem and Burlington, New Jersey, but not Philadelphia.

In an early land survey made by Thomas Sharp on Newton Creek, then in Old Gloucester County, mention was also made of an Indian field.

This twelve-year-old girl, in Indian costume, is a descendant of Samuel Bacon, the noted Quaker, who bought a tract of land of Shawkanum
and Et Hoe, Indian brothers of Cohansey Creek, 25th, 4 mo., 1683. Her
name is Edith Kirby.